Engineering is Broken

Ignacio Chechile

Engineering is Broken

We Are Doing It the Wrong Way

Springer

Ignacio Chechile
Helsinki, Finland

ISBN 978-3-032-18168-8 ISBN 978-3-032-18169-5 (eBook)
https://doi.org/10.1007/978-3-032-18169-5

This Springer imprint is published by the registered company Springer Nature Switzerland AG
The registered company address is: Gewerbestrasse 11, 6330 Cham, Switzerland

In memory of "El Bepi".

Preface

I've been closely watching how we do engineering for over two decades now. Not from some protected bubble or keynote speaker stage, but from the foxholes, knees deep in Stack Overflow threads, buggy code, paperwork, oscilloscopes, cables, and breadboards. I have been part of projects that made headlines for the right reasons and others that made headlines for the wrong ones. I have worked with geniuses who could solve impossible problems in the blink of an eye, and with people so out of their depth that it was painful watching them stumble with the most basic things.

Everyone I've met has taught me something; some by example, others by counterexample. I have seen how good design emerges from curiosity, collaboration, and iteration. I have seen systems born on humble whiteboard sketches and made a company worth billions, as I have seen overengineered approaches that didn't survive contact with reality. There is some irony in the fact that this book wouldn't have been written if engineering were perfect, but hey, medicine books wouldn't have a reason to exist without diseases!

The modern engineering workplace, if you look at it closely, is a living experiment in constant evolution. We have our rituals, our symbols, we speak in acronyms. In our quest to find the engineering rainbow, we work hard to create efficient machines while we neglect our own efficiency as the remarkable biological machines we are.

I believe we should look a bit closer at the things we create and learn from them.

We design systems that optimize throughput and speed, while our own throughput as engineers crumbles under the weight of red tape and nonsense. We add measures in our designs to make them reliable, while we love to create team structures abundant with single points of failure, as we pray for *that* guy to never quit.

Maybe that's the paradox of being an engineer: having the power to design performant beauty while being unable to embody it.

However, every now and then, even despite ourselves, we hit the right note. A prototype works. An elusive bug is found. A team jells, and for a brief moment, the noise fades, and you remember why you fell in love with this profession in the first place.

This book is about those moments—the good, the bad, and the ugly—and all the nonsense surrounding them. About how engineering is a constant fight against entropy, even if we spend half our time contributing to increasing it.

Helsinki, Finland

Ignacio Chechile

Acknowledgments The chapters in this book are shaped by my own experiences but enriched by countless stories I have collected from frustrated fellow engineers in innumerable casual conversations during lunch breaks, dinners, chats by coffee machines, and messages where people would open up and share with me the occasional nonsense of the day. I am immensely grateful for their openness, for it has provided me with invaluable insight about the dynamics of the engineering workplace.

Gossip has always had a bad reputation in the workplace. It's perceived as unprofessional. However, you just cannot ban gossip. Fighting gossip means fighting the very network that connects across the organizational fabric. In any workplace, there is the org chart showing the "official" network, and then there is the informal network, which is the actual nervous system of the organization. This network is simply made of people liking and trusting other people and talking frankly and safely about whatever they want. Formed on affinity, convenience, or a combination of both, these informal networks are ubiquitous, empowering, and evolve in real time. A stare in a hallway, an instant message, a conversation while queuing for lunch, information spreads fast. Gossip can help vent, offering a cathartic release from anger, guilt, anxiety, or some other unpleasant internal state, and a potential return path to a balanced state of mind.

Granted, too much gossip can be a symptom of an organizational issue. At the end of the day, gossip is like the body temperature of the organization: if it gets too high (fever), the problem is not that; something else is causing the temperature to rise. Companies use different mechanisms like bonuses, company-logoed thermos, promotions, and perks as organizational "aspirin"; they might lower the fever for a while, but it'll come back if the underlying condition remains.

Gossip networks are the nurseries where the resistance to organizational vices breeds, such as neglect, psychological unsafety, harassment, discrimination, and cronyism. If anything of that is happening, people are surely talking about it. I feel lucky that some colleagues chose to talk about those things with me and made me part, at least for a moment, of their quiet informal networks.

Competing Interests The author has no competing interests to declare that are relevant to the content of this manuscript.

Contents

1 Introduction 1
1.1 Structure of the Book 4

2 The Iceberg Theory of Engineering Management 7
2.1 Stupidity Creates Jobs 10
2.2 Stopping Stupidity 11

3 Excellence and Cynicism in Engineering 13

4 Fads and Fashions in Engineering 17
4.1 Is Engineering Becoming More "Digital"? 20
4.2 Digital Twins, Simulation, and the Price of Agency 21
4.3 Hype Me Like One of Your French Girls 25

5 The Velocity Fallacy 29
5.1 Do It Now 31

6 Design Versus Engineering 35

7 The Engineer Versus the Engineering Manager 39

8 Surviving Cringe 43

9 The Pretend Work Era 45

10 Prototypes and Products 47
10.1 You're the Real MVP (Minimum Viable Product) 48
10.2 Productize This 50
10.3 From Skunkworks to Production, and Back 51
10.4 Sticky Designs 52

11 Fake Requirements 57

12 Bad Work Breakdowns 61
12.1 Confusing Parts and Wholes 63
12.2 The Tree of Truth (Product Ontologies) 64

13 Fake Ownership 69

14 Minimal Engineering 73
14.1 Minimalism in Engineering 75
14.2 Intuitiveness and Idiot-Proofness 76
14.3 Nordic Moderation and Goldilocks Engineering 77

15 Shitty Tools 79
15.1 The Enterprise Software Shitshow 82

16 Ballooning Scopes 85

17 Death by a Thousand Pages 89
17.1 The Interoperability Paradox 91
17.2 Thirty-Thousand Requirements to Rule Them All 92

18 Engineering Myopia 97
18.1 Seeing the Crumbs 98

19 Appetite for Destruction 101

20 The Cold War Between Sales and Engineering 105
20.1 Product Management to the Rescue 106

21 Project, Systems and Products 111

22 Dangerous Originality 115

23 The Strange Immortality of Failed Projects 119
23.1 The Elves and the Project Manager 121

24 The Job Title Bingo 125

25 The Kindergartization of Work 129

26 Let It Rot 133

27 The Recurring Cast 135
27.1 The Floaters 135
27.2 The Assholes 136
27.2.1 The Making of an Asshole 138
27.2.2 Can It Get Any Worse? Yes, Assholes Might Team Up 140
27.2.3 Leadership by Erasure 140
27.2.4 Post-Asshole Life 141

27.3 The Wannabes ... 141
27.4 The Rainmakers ... 142
27.5 The Crony ... 144
27.6 The Tightrope Walkers ... 146
27.7 The Technocrat ... 146
27.8 The Anti-skilled ... 147
27.9 The Significant Other ... 148
27.10 The Single Point of Failure ... 149
27.11 The Delusional ... 150

28 Unrealistic Goals and Reality Distortion Fields ... 155
28.1 Setting and Adjusting Goals ... 156

29 Residual Taylorism in Engineering Design ... 159
29.1 The Ethereal "Processes" ... 163
29.2 Out-of-Context Efficiency ... 164

30 And Yet It Works ... 167
30.1 Near Misses ... 170

31 The Office Conga Line ... 173

32 Projects, Damn Lies, and Statistics ... 179
32.1 Double Schedule, Double Trouble ... 183

33 Stop Making Idiots Rich ... 187

34 What the Future Holds: Engineering, AI, and the Machine That Designs the Machine ... 189
34.1 A Robot Stole My Job, and I Couldn't Be Happier ... 192

About the Author

Ignacio Chechile is an engineer based in Helsinki, Finland, with more than 23 years of experience developing systems of many different sizes and complexities. His professional life pushed him into software quite early in his career, where he wrote hundreds of thousands of lines of code, many of which went to space. Later in his career, he moved into Systems Engineering and engineering consulting positions to the judgmental look of the Electrical Engineer that (still) lives inside him. Ignacio is an Expert Systems Engineering Professional (ESEP), certified by the International Council on Systems Engineering (INCOSE). Views in this book are strictly his own.

Chapter 1
Introduction

As young engineers, we dream of creating innovative technology that will transform the world. During our studies, we learn the fundamentals without talking too much about the day in the life of an engineer in the workplace. They don't teach us about office politics, cubicle life, or psychological safety. Then, with all the theory in our brains, we get our first jobs with excitement, only to find that real-life engineering is not so much about differential equations and algorithmic efficiency but about navigating social complexities, about compliance, bureaucracy, and run-of-the-mill activities. Soon, we realize that doing engineering feels more like hurdling than a professional practice. As we jump hurdle after hurdle, the thrill of doing engineering slowly starts to fade away.

And that, my fellow engineer, we must not let happen.

It is an unsettling truth that engineering, a discipline tightly coupled to modernity, is also one of the least scrutinized. We are surrounded by complex systems and objects designed and constructed by engineers. However, the inner logic of engineering practice is rarely subjected to much examination. At a time when the meaning of engineering seems up for grabs, with armchair experts and neophytes claiming to know how it's done, this book begins with a simple observation: while the world around us has changed substantially, the critique of how we engineer it has not.

The most incisive critiques of engineering practice—obligatory reads by Henry Petroski, Billy Koen, Archibald Putt, Donald Schön, or even the more sociological observations from DeMarco and Bob Sutton—are now several years old. Their insights remain very relevant, which is a confirmation of how little progress the discipline has made in confronting its inner demons. These authors skillfully highlighted the messy nature of design, the cognitive traps, the assholery that hides around every office corner, and the cultural blind spots of engineering organizations. These are big names, and by no means am I trying to find myself a spot in the altar of the great engineering management authors. I am only contributing with an update that some of the issues that have been there are still there, and also that there are new ones nurtured by new technologies and new social dynamics.

I. Chechile, *Engineering is Broken*, https://doi.org/10.1007/978-3-032-18169-5_1

Today's engineering landscape has been infected with smoke-and-mirrors, fashions, frameworks, certifications, overlapping standards, model-based-everything, and a sea of acronyms that more than help engineers communicate, build barriers between them.

The modern critique of engineering is fragmented, scattered, and lives in blog posts, Slack messages, and silent eye contact in meetings.

As I walk around the building of an engineering school on a university campus, there are testimonials from alumni hanging on the walls, I understand with the (good) intention of showing students some nice success stories and motivating them to graduate. One in particular catches my attention, which tells the story of a civil engineer:

> ...after graduating with a doctorate in civil engineering, [NAME] switched fields to business and now works as a management consultant with [BIG3 company] in New York.

I feel this testimonial nicely captures part of my thesis; we (engineering students, professors, practitioners) seem to have reached the point where we think this is fine. Don't get me wrong, there's nothing inherently wrong with choosing whatever career path you want; the wrong part is to consider this an "inspiring story" to motivate the next generation of engineers.

Ok, but where do we go to discuss the way we are doing what we do? Academia treats engineering more or less as a solved problem, probably because things *seem* to work. However, things that appear to work on the surface might be full of hidden flaws within and only work marginally, as we will see. Industry, on the other hand, treats engineering as a cost center, a resource to estimate and manage with spreadsheets. We have more tools than ever before; with their intricate subscription schemes and shiny user interfaces, however, they seem to be adding more fuel to the fire.

Now, some disclaimers.

First, this is not a sociological treatise. I am no sociologist or pretend to be. I am simply an engineer and accidental engineering manager who took the opportunity to observe and document, deep from the trenches, how our profession has submitted to the overall stupidity that has engulfed the workplace. Is it a rant, then? I hope it doesn't come across as one since rants are fueled by anger, and all I have in my heart is love (and some cholesterol). But if it does come across as a massive rant, so be it, I guess. One thing is for sure: this is not an anti-workplace, anti-corporate, *Fight Club-esque* manifesto kind of thing. We all need to work, and we haven't found, for now, any other way to do it without corporations, insecure CEOs, cubicles, accountants, and HR departments. In any case, just make sure you don't take this book too seriously, and don't feel offended if you see yourself in the picture and you don't like it. As Ricky Gervais said during a Golden Globe awards ceremony: we're all going to die soon, and there's no sequel.

Second, this is a book with no pictures. Have you ever tried to get a professional illustrator to pick up your calls? They're busy people. Also, there's an overuse of stock photos in books, with the classic ones of people touching screens floating in midair, waterfalling source code, people wearing virtual reality goggles, and robots doing weird things. Lots of stock photos of people smiling at technology? Why do they smile to a laptop? Anyway, I trust you can do your own Googling if you need a visual representation of anything I say here. Books try to keep you hooked and increase their page count with pictures and visual niceties. I don't. Go and find the visual cues you need, and please consider coming back. This will be a test of your attention span. If you never come back, maybe our thing was never meant to be. Also, no contrived block diagrams to make myself look smart, no overused comics to make myself look funny through the work of others. I have purposely avoided links in footnotes as they go broken over time, but also to respect and protect boomers who might be reading this in physical format and might try to push the links insistently with their fingers until they bleed.

Third. AI is a hot topic at the moment. Similar to the side-looks calculators would get back in the day, generative AI is frowned upon when it comes to content creation. Just like no one would give you a bad look today if you can't mentally multiply 813 times 234, the time will come when AI will be perfectly coupled to every creative process. Still, it's a good moment to disclose that I have used your usual AI tools not to create textual content but as an adversarial editor to help me challenge some of the claims I am making here, mainly to avoid sounding inflammatory without substance. Have in mind that English is not my first language, so I have run passages through grammar correctors to get it right, or at least to get it better. Most chapters in this book have existed for a long time as standalone articles in a couple of blogs I used to maintain. This book is a curated collection of those, all interconnected by an underlying story: the story of how we are doing engineering the wrong way, and how life is too short for spending so much energy on the wrong things.

Fourth, I am aware that the title talks of "engineering" in a very broad sense, and engineering is a rich ecosystem of different kinds and subtypes. Most of the disorders and peculiarities I will talk about in this book apply mostly to the branch that deals with software-intensive and cyber-physical systems, domains where many different engineering specialities converge. A civil engineer working on building bridges or extending railroads might not feel totally related to some of the oddities I describe here, since their practice might not have yet been hijacked by nonsense, or at least not as much. To those, what can I say; you don't know how lucky you are! Enjoy while it lasts and use this book as a little window to a future that can materialize in your domain if you don't do something about it.

Fifth, I am not writing this from a standpoint of "you're all doing it wrong, and I know how to do it right". I have made countless mistakes both as an engineer and as a manager, many of which still live rent-free in my head. And I am probably not done screwing up. I am not sure I know what "good engineering" entails in absolute terms because it might be context-dependent. But I know what bad engineering is. I know it when I see it. You know it when you see it.

Sixth and last, this is not a book of magical solutions. The only pro-tip I can provide to you it's not something I have discovered, but something that it's been there all along in front of your eyes, and it's open-source and free: the most underrated life hack is doing the damn work.

1.1 Structure of the Book

I have written this book with the intention that you can do some catharsis with it. I like to imagine this situation when you leave yet another meeting when something incredibly stupid has been thrown on the table. I like to envision that, right after all that nonsense, you come back to your desk, sit down, open this book, go through the index, and have a soothing moment finding that same stupidity described here, feeling a little bit less alone in the process. If this book can act as a valve and give you some company in your journey as an engineer, a big part of my mission will be accomplished. I am sure I haven't covered all possible scenarios and situations, so I apologize beforehand if the catharsis is not complete. Every chapter in the book is rather self-contained, so you can jump around depending on what the nonsense is in order at your workplace on a particular day.

When stupidity kicks in, try to relax, make yourself your favorite beverage, and remember that engineering is more than what the occasional corporate nutcase wants you to think it is. You know what it is. You worked your ass off to become an engineer; those fools can't get under your skin. And remember that you can't fight the inescapable Putt's law,[1,2] which observes that technology is dominated by two types of people, those who understand what they do not manage and those who manage what they do not understand.

---

[1] *Archibald Putt* is the pseudonymous author of *Putt's Law and the Successful Technocrat: How to Win in the Information Age* (Wiley, 2006), a compendium of essays about technology management. The true identity of Putt has never been publicly confirmed.

[2] Note that the impact of Putt's Law can be delayed, but the Law itself is never broken.

Chapter 2
The Iceberg Theory of Engineering Management

Abstract Overconfidence kills, figuratively and literally. Have you ever felt like the bosses seem to be constantly steering the boat into collision course with their half-baked decisions and ridiculous trains of thought? Are they suicidal, stupid, or maybe both? Rest assured that they know what they're doing. In this chapter, we dive deep into what makes them behave that way and brainstorm what it may take to make it stop.

Keywords Overconfidence · Decision-making · Organizational stupidity · Functional stupidity · Engineering management

In 1996, Alan Sokal wrote a paper obscurely titled "*Transgressing the Boundaries: Towards a Transformative Hermeneutics of Quantum Gravity*", and submitted it to *Social Text*, an academic journal of cultural studies published by Duke University Press. The paper would eventually get accepted for publication, and Sokal's life would never be the same. A few weeks later, Sokal revealed in another magazine that the paper was a hoax. Utter rubbish. But a carefully-crafted, intentional rubbish. Why? Sokal wanted to prove that postmodern philosophy was basically full of it and rife with a pompous and inaccurate use of mathematics and physics, precisely what his paper was replete with. And, boy, did he succeed in proving it.

Sokal would then publish two derivative books: the absolutely great *Fashionable Nonsense*,[1] where he and his co-author Jean Bricmont analyze in depth texts from renowned authors like Lacan, Julia Kristeva, and other encumbered intellectuals and show in detail their incompetent and pretentious use of scientific concepts, and *Beyond the Hoax*, a detailed recollection of how the hoax was conceived and executed.

[1] *Fashionable Nonsense: Postmodern Intellectuals' Abuse of Science*. Alan Sokal & Jean Bricmont. Picador (1998). The book has hilarious passages; I strongly recommend the part about Lacan's use of imaginary numbers.

I. Chechile, *Engineering is Broken*, https://doi.org/10.1007/978-3-032-18169-5_2

Ultimately, the Sokal hoax revealed how the evergreen "envy of physics[2]" was and is still very prevalent, permeating the humanities and social sciences across the board. With Alan Sokal's work vigilantly checking on me, I will use some analogies in this chapter that use physical concepts like forces and buoyancy to illustrate social constructs like organizations. Be sure that I am fully aware that there is no real fluid dynamics involved here, except for the liters of coffee that were consumed while writing it.

But first, some context this is my own observational theory, and perhaps it's grown from my burning need to understand why some tech managers I have worked with behaved the way they did.

The theory suggests managers are in a constant—albeit probably unintentional—pursuit of *sinking the ship*, which in the end doesn't sink because of competent engineers in the organization acting as the counterforce that prevents disaster. A part of me wants to think this damaging behavior is not intentional because they are on the same boat that they're actively trying to sink. Some further analysis reveals that managers intimately know they can count on the self-preservation instincts of competent people around them to always steer the boat back to safety; this way, they can afford to "rock it[3]" the way they do. Edward Smith, the famous captain of the Titanic, once said:

> When anyone asks me how I can best describe my experience in nearly 40 years at sea, I merely say: uneventful. I never saw a wreck and have never wrecked, nor was I ever in any predicament that threatened to end in disaster of any sort.

The interesting thing with Mr. Smith is that the Titanic wasn't his first blunder, only his last. Captain Smith once rammed the RMS Olympic into a Royal Navy warship, the HMS Hawke. After each collision, they kept promoting him. A dynamic that sounds familiar, doesn't it[4]? Maybe that previous quote is apocryphal, as some sources claim. But Captain Smith—this time 100% verified—did say about the Titanic in an interview with the New York Times in early April 1912:

> I cannot conceive of any vital disaster happening to this vessel. Modern shipbuilding has gone beyond that

Now, one doesn't want to dunk too hard on Mr. Smith. After all, the scale of the Titanic tragedy isn't exactly something to laugh about. However, something can be stated objectively: as captain of the ship, responsible for delivering thousands of lives safely to their destination, the man was plethoric with confidence. Armed with that unshakable self-assurance and unstoppable optimism, he managed to plow into an

---

[2] The phrase "envy of physics" is used to describe the tendency to admire and sometimes imitate the elegance, precision, and predictive power associated with physics. It reflects both a real aspiration for rigor but also a frustration: physics can often express complex realities with a few compact equations, whereas many other fields deal with systems too messy, contingent, or context-dependent to be reduced in the same way.

[3] The Third Anomaly of Putt's law tells us that a person rocks the boat to get ahead in a hierarchy.

[4] The evergreen Law of Failure, which states that organizations punish small failures but reward big ones.

iceberg in the cold, vast North Atlantic and secure his place in history as the man who sank what was thought to be unsinkable.

What's my point here?

Sometimes it seems that the "captains" of the metaphorical corporate ships we sail on share that same absurd excess of confidence, neglecting the surrounding signals and bringing us every day a millimeter closer to the iceberg. With half-baked decisions and methodologies imported from contexts that have nothing to do with ours, with convoluted and abstract "processes" that make everything harder, with mediocre ideas pulled out of thin air without a shred of analysis of their unintended consequences, with random and unnecessary changes. They keep putting us on a collision course.

So why don't we constantly end up like Rose on that floating door in the award-winning, painfully-long movie from 1997?

Because of the quiet, thankless, continuous work of unsung heroes sitting in the lower ranks of the organization, who tirelessly compensate for the blunders of decision-makers and thus avert disaster.

While the ones with power are busy cooking their next "brilliant" move, the informal networks quietly fix things, while rarely getting any credit for it.

But wait, it's even worse. If you think your captain calmly and constantly sailing you to a shipwreck motored by overconfidence is not worrying enough, there are more "forces" acting on the ship, one pushing up, and another one pushing down. As for the latter, Mats Alvesson and André Spicer call this organizational stupidity, which is the proliferation of practices and behaviors that add no value but, *au contraire*, consume energy and effort from many. Fads, bureaucracy, vapid methodologies, and an unnerving urge to maintain the status quo. They elaborate:

> [Stupidity] can become self-reinforcing. This happens when employees stop asking searching questions and are rewarded with a sense of (false) certainty. It happens when they are good team players, reliable followers and well adapted organizational members who do not threaten their managers or colleagues. It can create an optimistic narrative. People tell good-news stories. Sometimes there are rewards such as promotions, pay rises and pats on the back from people who matter. All this is taken as evidence that things are going well, and you should continue to follow the same course.

Organizational stupidity increases with headcount, and it is practically an inevitable result of growth itself. With more people, more bullshit positions, and more managers equipped with a breadth of insecurities, the organization eventually turns in on itself and not only stops doing meaningful work but engages in damaging practices.

Now, in the opposite direction, the buoyant force keeping you afloat is, as we said, competent people and their self-preservation instincts, but also having a good product. Found either by luck, serendipity, or vision, it doesn't matter; a good product can pull you upwards with the force of a thousand dolphins on steroids.

Take any blue-chip company. They all have tens, if not hundreds of thousands of employees worldwide. It is hard to believe those corporations are not rife with drama, politics, egos, and thoughtless practices across the board. Feels almost natural that organizations of that size would grow such a scene. But they can afford to let some stupidity run free because they are resourceful, known, and most have mature products and solid revenue streams that give them reliable buoyancy.

Stupidity—the sinking force—does increase with growth. Then why on earth would anyone grow too big before having a good product? That sounds like basically loading bags of bricks on the boat and going for a sail. If you do not have a good product, the last thing you want to do is to onboard a massive slab only to sink like a stone in a pond. And yet, many do. Why? Well, investors' capital tends to mask the need for buoyancy, by providing a virtual force up. By accessing easy money, the hiring spree kicks in—and usually targets irrelevant, non-product-related positions—and when the resulting sinking force out of all that complexity is already too strong, investors start asking questions, the rug gets pulled, and it is too late to think about coming up with a plan.

However, we have seen big organizations surviving with absolutely no product, haven't we? How do they manage to live? Simple: Some manage to craft a lifeline that keeps them artificially alive for long times, like a ventilator. It is typically subsidies, investors' money, governmental and institutional funds, or an explosive combo of all that.

Great products can emerge from surprisingly small resources, and perhaps they're best born in tight, scrappy circles. The iconic personal computer company from Cupertino famously started with just two friends in a garage, long before they became symbols of corporate power. Back then, they weren't giants; they simply saw the edge and went for it. First the product, then the rest. Today, the same company can easily afford all sorts of internal drama—which they have— because its buoyancy is so great it practically floats on saltwater.

Companies do not succeed because they get rid of stupid practices. They cannot do that and will never be able to do that as long as they continue employing human beings. Organizational stupidity is inherently human. The reason why they succeed is that they use the energy and gray matter at the right time to figure out what the goddamn product is.

2.1 Stupidity Creates Jobs

How many times have you encountered someone in your organization who wouldn't have a job if things were done the right way? There are countless people employed exclusively because things are done poorly. They're hired to fix messes, but they never fully fix them because doing so would eliminate the reason for their existence. In the startup world in particular, where growth is often seen as a figure of merit by investors who love looking at cold metrics, having a lot to fix is a great excuse to open job positions and look busy, even if you are busy with the wrong things. Incompetence creates millions of jobs worldwide every year and shows no signs of slowing down, pumping up the global GDP.

We have built broken systems and organizations in the past, and nothing suggests we'll stop building them in the future, and while trying to repair them we make them worse, which creates even more jobs. Of course, the wheel meets a limit: if the organization goes too stupid, you cross the line when it can't be fixed anymore, and hiring firefighters won't help. Nothing will help once you cross the "unsalvageable stupid" point.

Systems and organizations decay just as an avocado does, primarily because the surrounding entropy inevitably increases over time, and they cannot maintain themselves unless energy is continuously used to adjust them as conditions evolve. For every functional arrangement that achieves a desired outcome, there are orders of magnitude more arrangements that accomplish nothing or in fact degrade performance. Every system ever designed must be actively maintained, and organizations cannot escape this truth; we must invest substantial energy to keep it in an ordered state. Leadership quality also decays (and quickly) if left open loop; if you let the bar go low by doing nothing as your leaders decay like bananas in the sun, just be ready for some suffering.

Organizational decay plants the seed for bureaucracy. Everyone loves to hate bureaucracy; it's even fashionable to complain about it. However, bureaucracy is an effect, not a cause. The memos, the documents, the pointless meetings, the paperwork, the bad tools, none of these are really to blame. They're the visible part of a broken organization down to its roots, which they can't fix by themselves.

2.2 Stopping Stupidity

Ok, all that being said: Is organizational stupidity even stoppable?

We all agree on the sheer level of idiocy that Homo sapiens can reach in its corporate form. Whether it's congenital stupidity, acquired stupidity, or stupidity by peer pressure seen from the outside, I do not know. But it's there. So, we must accept that stupidity will always exist. As someone said, an idiot is born every minute. We can't dream of a zero-stupidity world where every idea and decision is a hymn to common sense.

So, who stops the stupidity, then?

There are two ways this can go. We can let stupidity flourish and bloom happily in an evergreen garden of dumb ideas, where nonsense constantly overflows from the top down, and we end up resigning ourselves to accept whatever "was decided from above", focusing only on minimizing the impact of the current bullshit on our miserable lives.

Or, we can fight it. But of course, the question is, who has the strength to fight such a strong force of nature?

Every organization needs a role that often goes unnoticed but is absolutely key for survival and collective mental health: the stupidity stopper. Also known as the nonsense blocker, the bullshit brake, the dream crusher, the soul calibrator, the delusion deflector. Someone with a clinical eye, a sharp mind, and a karate tongue. A born ankle-breaker of absurdities, always ready to defuse the dumb initiative of the day, send the "visionaries" back where they came from, and tell them that what they just proposed is, quite simply, a load of crap. It's a close sibling of the Technocrat, which we describe a bit further.

Needless to say, it's not an easy role. The stupidity stopper needs backbone, credentials, natural authority, and a good dose of common sense to earn that position, a position that's far from pleasant, since there's always the risk of being labeled as the one who blocks everything. Of course, the stupidity stopper runs the risk of getting drunk on power and turning into a staller, whose only job is to tell everything that moves to "get in line" while making sure nothing ever gets done.

Furthermore, stupidity is multidisciplinary. There are technical stupidities, sales stupidities, HR stupidities, and legal stupidities. You can't expect to have a polymath who can block every single one from every possible domain. Ideally, you would have a stupidity stopper per domain or area.

And then there's scale again. Stopping nonsense in a company of 50 people is one thing; trying to do the same in one with 1,000 is a whole different song. In organizations of that size, mini-empires emerge, fueled by spurious alliances, nepotism, and other organizational vices that inevitably come with scale. At that point, no amount of backbone or credentials can save you: a single soul can't possibly keep up with the sheer stupidity per second being generated. Forming a compact team of experienced, legit, no-nonsense people—a sort of Ghostbusters of bullshit—is paramount.

Without anyone to block it, zap it with a proton cannon, or kick it where it hurts, there's nothing left to do but sit and watch as stupidity grows like ivy and takes all our walls.

Chapter 3
Excellence and Cynicism in Engineering

Abstract Do we still care about excellence in Engineering? What does it mean in reality? The saying goes that excellence is doing ordinary things extraordinarily well. Our craft has evolved in a way that, instead of pursuing excellence and improving what doesn't seem to work, engineers drift into cynicism, where they complain against everything and everyone. How do we reset from all that and find thrill and beauty in the ordinary again?

Keywords Excellence · Cynicism · Engineering management · Professional improvement

One of my mentors, but more importantly a good friend this profession has given me (P.G.), kindly sent me an old book from his personal collection, titled "Excellence in Engineering" by W.H. Roadstrum (Wiley, 1967). Roadstrum's work goes to a great extent to define an engineering activity that seeks to project its practice away from mediocrity. According to Roadstrum himself, his book was inspired by another book called "Excellence" by John W. Gardner, which dives into analyzing the relationship between excellence, education, and equality. One phrase from Gardner's work has become quite famous: "Excellence is doing ordinary things extraordinarily well". Another part goes: "An excellent plumber is infinitely more admirable than an incompetent philosopher. The society that scorns excellence in plumbing because plumbing is a humble activity and tolerates shoddiness in philosophy because it is an exalted activity, will have neither good plumbing nor good philosophy. Neither its pipes nor its theories will hold water".

Engineering is a craft, as plumbing is. Just more glorified. Such glorification can blind us from pursuing the better. A pursuit which should never stop nor relent. Being excellent means delivering things that work and work well and, in the process, deriving pride and satisfaction from the work done.

Excellence is an endeavor, not an end goal, and it spurs us to wake up each day to perform better and become better at our craft, no matter how ordinary.

Some sharp readers could argue that it's office-based work that is broken and not engineering. They could argue that we haven't found a way of working collectively

I. Chechile, *Engineering is Broken*, https://doi.org/10.1007/978-3-032-18169-5_3

without falling into traps and silly practices, be it finances, healthcare, consulting, or engineering. And there is some truth in that analysis: we haven't. However, engineering has let the stupidity filter in with too little resistance. And here we are.

I have seen fellow engineers grow so cynical that they criticize their jobs every day, no matter what. Bad decisions from corporate suck, but also good ones. And that truly makes me think: if you hate it, then why stay? Sure, we all have bills to pay and mouths to feed—there ain't nothing in this world for free—and changing jobs is not easy, but I sense part of that cynicism is nurtured a little bit every day when vapidity gets in our way, when we are not allowed to do engineering the way we were taught to. When the noise is stronger than the signal. But here's the thing: I strongly believe that we as engineers are responsible for the quality of our practice. Who else will do it for us? We are the ones responsible for lowering the surrounding noise. Yes, the bosses have the power and may call the shots, but engineering departments are the foundation of every organization that sells technical products. So, we have the power, and we must use it accordingly (and responsibly). Putting our heads down and accepting things that affect our practice only makes it worse. Complaining next to the water dispenser with your colleague might help you vent, but it does nothing for the surrounding stupidity to go away.

Plato, in his *Apology for the life of Socrates*, reminds us that all societies need a "gadfly" of sorts to sting the "steed" (In Plato's work, the state) into acknowledging its proper duties and obligations. Societies in all their shapes and sizes need those who can question the state of affairs.

Uncritical acceptance of the status quo is a big problem. Just like in the famous 5 monkeys' experiment and their "don't ask me, that's just how it's always been done", lying back in eternal dogmatic slumber is harmful for the collective because then, who's there to call out half-assed ideas from spreading like mold? Who stings the steed?

However, I am aware there is a thin line between constructive criticism and cynicism.

The modern version of cynicism is considerably less elevated than its Greek precursor. Modern cynics are those who always mistrust everything and everyone. Continuously flagging flaws and yet doing nothing to solve them, they are expert problem spotters while largely uninterested in finding a solution. Note that cynicism is not the same as pessimism. A pessimist is capable of trusting, although they always expect unfavorable outcomes; a cynic trusts nothing and no one.

Cynics blame the *system* for their disappointments and failings. There are cynics in almost every organization, and their spirits surface as the organization grows, as promotions take place, and as hierarchies and policies appear. As the *system* emerges. For a cynic, as soon as such a *system* is perceived, the system must be fought. They intimately need the *system* to stay imperfect, so it can continue fueling their discontent.

While reading this text, some might arrive at the conclusion that I might be just another anti-system cynic, criticizing everything from a safe distance, with no skin in the game.

I guess a tad of cynicism does live in me. I do loathe elitism, nepotism, and cronyism. I enjoy scrutinizing and criticizing those calling the shots because I am convinced it's a valid mechanism to keep them accountable for the immense power and influence that they can exert on their subordinates. I enjoy *stinging the steed*, without implying that I think I'm Plato.

However, I try to balance this cynical side of me with a dose of anti-cynicism. Unlike full-time cynics, I don't fight *the system*. I'm positive *the system* can be made better. And if I feel I can't bend its odds, I have always chosen to move on instead of staying just to complain next to the coffee machine.

We need to understand what our role is as engineers in keeping our profession true to its foundational values. Whether we still remember what "better" means when there is so much surrounding noise. Nobody else will do it for us.

Chapter 4
Fads and Fashions in Engineering

Abstract Fashion doesn't only shape physical objects or garments. It also affects the way we work. And that includes the way we do engineering. The history of technology is too often told as a story of rational progress. In reality, it can also be told as a story of fads. Engineers think of themselves as immune to cultural forces, but no one builds inside a bubble. We design and build inside waves of imitation and cultural trends that shape everything. Agile, vibe coding, and whatnot, understanding fads does not necessarily mean dissing them, but recognizing the behavioral patterns underneath the surface helps us better choose what makes sense and ditch what doesn't.

Keywords Fads · Fashions · Trends · Habituation · Hype · Gartner cycle · Agile · Vibe coding · Imitation · Design · Digital engineering · Digital transformation

The history of technology is too often told as a story of rational progress. In reality, it can also be told as a story of fads. Fads not only appear in clothing or consumer taste, but also in other domains like cars, design philosophies, and engineering culture.

Fads are transient waves of collective enthusiasm that shape markets, careers, practices, and the direction of innovation itself. They appear seemingly out of nowhere, rise fast, and once obsolete, they leave behind a trail of vintage jargon, and abandoned practices that look ugly and bizarre when seen from the distance. To understand fads, it is necessary to step out of purely technical frames of reference and look at the psychology and sociology that lie underneath it all.

Fashion has always existed, but industrialization took it to new levels. In medieval times, only aristocrats could afford novelty. By the nineteenth century, textile mills and railways turned fashion into a mass phenomenon. By the mid-twentieth century, cars, appliances, and garments were changing every season. General Motors perfected the art of planned obsolescence with annual model refreshes with slightly new chrome, fins, dashboards, and slogans. Function remained roughly constant: 4 wheels rolling to haul your ass from A to B, but aesthetics became the cow to milk to keep on making money out of underlying technology that was not changing fast enough to keep revenue flowing. Automotive design is a great example of

I. Chechile, *Engineering is Broken*, https://doi.org/10.1007/978-3-032-18169-5_4

fashion waves. Tailfins in the 1950s, the boxy austerity in the 1970s, the (somewhat phallic) aerodynamic obsession in the 1990s (including that hideous wooden-like era of interiors, yack), and the EV minimalism of today. These shifts are not driven by breakthroughs in automotive engineering at all, but mainly by cultural waves boosted by marketing and consumer trends. The underlying engineering changes little between fad cycles: what changes is what the machine represents. A 1959 Cadillac tailfin was never too useful. A 2020s electric sedan's spartan dashboard with a single OLED screen flags modernity and the win of the digital over the physical; one over-the-air software update and you get a new dashboard, if you don't brick it in the process.[1]

These waves are shaped by a complex mix of social forces, novelty seeking, and human desire. Once a form factor is associated with status or with the future, it spreads through imitation, only to decline when overexposed.

Fashion doesn't only shape physical objects or clothes. It also affects the way we do things. And that includes engineering. Software engineering is notorious for its succeptibility to fashions. The 1990s saw the rise of formal modeling and object orientation; in the 2000s, Agile came to save us from the scary waterfall, and then morphed into a kind of fundamentalist religion. Scrum boards, standups, and sprints didn't precisely spread because of their effectiveness but because they irradiate adherence to the cause. Any random team can point to the sticky notes on their walls and declare themselves agile, because who will come and say otherwise? There isn't an Agile police. Then came Lean, Kanban, DevOps, NoOps, Serverless, microservices, monorepos, centralized architecture, decentralized architecture, move fast and break things, move fast but don't break things, failure is not an option, failure is an option, 10x engineers, everything-as-a-service, vibe coding, and whatever label happens to dominate the current moment. Most of these waves contain a seed of legitimate insight, but they become fads when the ritual survives after most of the meaning is gone. Just as a tailfin doesn't make a car faster, an issue-tracking board doesn't make a team engineer better.

The cycle is predictable. A practice appears to solve a (real) bottleneck in a specific context. Influential companies adopt it; it becomes a marker of being "ahead of the curve". Books, frameworks, certifications, and trainings emerge. Everyone uses it, often without understanding or questioning why. Eventually, it loses its halo, habituation fatigue kicks in, and a new fad rises.

The sunsetting of a fad happens through low-key obsolescence, that is, without fanfare. A methodology is tacitly declared legacy when it no longer provides a "cutting-edge" identity. Teams adopt the new fad to avoid the stigma of being behind. Mind you, the mechanism is lucrative: it feeds entire ecosystems, including armies of consultants, tool vendors, dubious conferences, and social media influencers. In cases, a fad might be pushed out of its original domain and used in areas quite dissimilar from where it originated. Today, we have Agile marketed for things way

[1] Recent software issues have caused cars to "brick" or become inoperable during software updates, including cases where cars required a technician to manually open the door and reset the computer before it could be towed to the dealership because the keys were no longer recognized.

beyond software: Agile communication, Agile Business Development, Agile families (yes, there are some people out there having retrospectives with their kids). All with their obligatory coaches cashing in, and the obligatory articles in renowned business quarterly magazines. Yuck.

But why things like vibe coding or Agile "click", whereas other things equally vapid, do not?

Certain fads succeed because they somehow resonate better with deeper psychological, cultural, and organizational incentives than others.

For a fad to click, it must be accessible to a broader and somewhat intellectually easy-going (read, lazy) audience: simple enough to communicate and imitate without deep expertise. "Agile" was never a very rigorous methodology, but it's its branding, it's its positive wording that takes it to the next level; the same thing with a different name such as *Customer-centric Iterative Software Engineering Approach* wouldn't have clicked the same way. The same happens with "vibe coding": it conveys a sophisticated posture without requiring a 500-page manual. Who wouldn't like to *vibe*?

Ideas that remain too niche die out: they often require a lot of context, deep technical grounding, and don't translate well into a slogan. Fads bloom on compactness. The easier you can shove it on a pitch deck, a LinkedIn profile or a CV, the faster it will spread.

Engineering fashions find their way in because they're loose enough to be reinterpreted to fit almost anything. They work because they're deliberately ill-defined: they can mean so many things that they actually mean very little. I just can't put my finger on it, but you feel almost as if there was some Barnum effect[2] here at play, making them a close sibling of the horoscope.

Fads are not totally pointless in the end. They remind us that technology is not as rational as we think. Anything engineered, from cars, software methodologies and frameworks, is both a technical artifact and a cultural object. Engineers like to imagine that superior solutions prevail through technical prowess alone. But that's wrong. From the Cadillac ridiculous tailfin to the agile "transformation", fads have shaped industries, employment, and even how we think about technology. Sometimes they drive useful change. Sometimes they waste lots of resources.

Engineers think of themselves as immune to cultural forces. But no one builds in a bubble. We design and build inside tides of meaning, imitation, and cultural trends that shape everything, including ourselves. To build things that work (and sell), you have to learn not only how to engineer but also how to navigate the human factors that guide the way we create technology. 50 years down the road, it will look and sound ugly and ridiculous anyway.

[2] The Barnum effect is a psychological phenomenon where people interpret vague, general statements as highly accurate descriptions of themselves. Named after showman P. T. Barnum, who famously said there's "an idiot born every minute", it explains why horoscopes, personality quizzes, and frameworks feel "spot on" even though they could apply to almost anyone. The key is that the statements are broad, flattering, and open to personal interpretation, allowing individuals to project their own meaning onto them.

4.1 Is Engineering Becoming More "Digital"?

Betteridge's law of headlines reminds us that any headline that ends in a question mark can be answered by the word 'no'. So, you start to get an idea about the answer to the question that titles this subsection. But hey, let's at least give it a chance and see where we stand.

Since we are discussing fads and fashions, it's relevant to bring to the center stage the blurry "digital engineering" and "digital transformation" concepts which are, at the time of writing, quite hot in quarterly business articles and management consulting firm websites. Everyone claims to be doing *digital engineering*. But are they really?

As it happens, the hype industry (more about this in the next section) has hijacked and bastardized the concept of "digitalization" or "digital transformation" until it means nothing, propelled by management consultancy firms that spout the obligatory low-effort/low-quality white papers with loads of filler and high-res stock imagery galore of people smiling to screens and no substance whatsoever.

This is quite sad if we consider that there is a need for a true, honest discussion on how our engineering work transitions from the computers into the not-well-behaved physical world.

Until very recently, detailed knowledge of a physical object required being physically near it. Information and object were essentially inseparable. Descriptions existed, but they were shallow and inconsistent, especially before the late 1800s, when basic agreements like the standard meter finally made interchangeable manufacturing plausible. Before that, people would use less consistent methods, such as a king's body parts to standardize units of measurement, like King Henry I of England decreeing the yard to be the distance from his nose to his outstretched thumb.

Only in the latter half of the twentieth century did we learn to extract information from objects and encode it digitally. Early CAD models were sparse, static sketches of pure geometry. Over time, they evolved into more dynamic counterparts, including digital representations that don't just show form but simulate behavior under forces and over time.

However, working solely with physical prototypes had inherent limits. Each required bespoke fabrication, dedicated test environments, and destructive tests. We could only explore the scenarios we anticipated, meaning the first encounter with an untested condition often happened in real use, with failures that could damage property, injure or kill. When these objects were combined into systems, the complexity grew exponentially, and the whole behaved in ways no collection of part tests could predict. Systems Engineering was born. As software became part of these systems, the need to simulate not only physical behavior but also computation became unavoidable. And with this shift, one challenge became clear: the fidelity of our simulators mattered a lot. The fidelity of simulations is at the center of challenges related to digital engineering, and we will dive a bit deeper into that soon.

The Digital Transformation heavily relies on the Digital Twin concept. The Digital Twin, in its definition, means to be able to design, evaluate, verify, and use or operate the virtual version of a system of interest during a large portion of the development lifecycle. In theory, with Digital Twins, we could understand whether our designs are actually manufacturable and determine performance and modes of failure, before a single bolt is fabricated. Digital twins also downplay the role of physical prototyping, as the need to produce instances of the design is replaced by analyses done in the digital domain.

In reality, there is an ugly side to the "Digital Twins" and the transformation they are supposed to bring with them. Far from the ideal of a cohesive, full digital replica, representing a complex system in the digital domain requires large amounts of unstructured information in the shape of disaggregated tools and software environments that do not easily talk with each other. This includes mechanical design data, thermal and fluid dynamics data, power systems, Multiphysics simulation, finite elements analysis, electromagnetics, application software source code and binaries, firmware code and binaries, and programmable logic, just to name a few. As much as the Product Lifecycle Management tool makers have tried so hard to sell to us that they can easily handle all that, at the time of writing, it all remains a wet dream. PDM/PLM environments only bite an infinitesimal portion of the problem, usually the easiest portion without complex runtime behavior.

The digitalization of engineering (and product design) is progressing, yes, but still in its early stages and looks more like an expression of desire than a real transformation. Today, it's mostly a disjointed collection of expensive tools that come with industrial amounts of vendor lock-in and are only affordable for those with really deep pockets.

For most of us mortals, the "digital transformation" it's still largely our beloved collection of Excel spreadsheets and an inelegant bag of scripts.

4.2 Digital Twins, Simulation, and the Price of Agency

Big consultancies love to propel the idea of the Digital Twin as the savior of all our problems. So, let's dissect what it takes to actually simulate a complex object.

When we design and build complex systems, we are reordering unstructured matter in a very specific way, with the objective that the new arrangement provides a function that the unordered pile of matter is not able to provide. Be it a Boeing 747 or a microchip, they are basically a finite number of atoms stuck together, arranged to perform a specific task. Same for a smartphone, a submarine, a bicycle, or a data center.

Now, the macro function of these piles of atoms emerges as the process of arranging them reaches certain completion stages. A microprocessor becomes able to act like one once the different building blocks materialize and can connect and interact. However, as we group matter into different sets of discrete parts, the emergent behavior of the collective starts to cloud the intrinsic agency of its constituent parts. This means that our awareness that objects are composed of smaller parts gently fades away. The beauty of complexity is that we can blissfully forget that the objects we design are made of a zillion tiny particles, so we can focus on the macro function that the object provides.

A simulation of a complex object is a virtual representation of all that—both structure and dynamics—in a computational way. While modeling, we go from coarse to fine, and we choose to focus on certain relevant macro-level functionalities and abstract ourselves from the reasons why this macro function emerges. For instance, when simulating a microprocessor reading or writing from memory, we do not need to simulate the electromagnetic wavefronts propagating through copper interconnects for a receiver to resolve if the data is a 1 or a 0, nor do we need to simulate charges moving across a transistor to replicate how digital logic works. When the level of abstraction is high enough, a memory read is just a value put in a position in memory. That's enough. Essentially, the transfer of data is reduced to function calls.

The reason for going coarse is optimization; more agency requires more computational resources allocated to compute each agent's state evolution. The more details we add to our models, the longer it takes to create the model, the more resources needed for the model to execute, and the more complex the model gets, which invalidates the idea of modeling in the first place; the "usefulness" of a model decreases as its fidelity increases since it becomes as difficult to understand and operate as the actual system. This is beautifully captured by Bonini's Paradox, which says:

> As a model of a complex system becomes more complete, it becomes less understandable.

This is somewhat related to Paul Valéry's famous phrase:

> Everything simple is false. Everything complex is unusable.

If the object we are trying to model has a hierarchical composition of components that are also composed of other components, then it means the physics and behavior of those collectives can be simpler to handle if we conceive and treat them as wholes: they move together, rotate together, and interact with the environment together. The constituent atoms and particles of those groups are interdependent enough to follow each other everywhere they go, which means we can only care for the group as a thing on its own and allocate computational resources for that. Often, what we model does not show such "chunky" unity. For example, things get more complicated when we have to model and simulate fluids or gases. We can't really model a fluid or a gas as one monolithic behavioral entity. Simulating chemical processes, for instance, processes like Atomic Layer Deposition (ALD) used in the semiconductor industry, demands an insane amount of computational resources because simulations must account for thousands of atom interactions, and each one of those is relevant for the final result. This is the domain of Computational Chemistry.

What price do we pay for neglecting the almost infinitely compound nature of objects when we model? Fidelity of behavior and accuracy. Our models will not behave as the real thing, nor provide the same results as the real thing: that's an inescapable fact. We might be ok with this, but only to some extent, and under certain conditions.

So let's see what it means to create a true digital twin. Consider a rudimentary electric circuit: a battery, a resistor, and a switch, which we happen to have as a real circuit on top of our lab table.

Imagine we also want to model the circuit in a simulated capacity. How do we proceed? Well, we can just do this in a spreadsheet and implement Ohm's law. Is that a model? It is, in a way, since it provides some insight into how different parameters of the circuit interact. In this spreadsheet model, of course, there is no real current, or real resistance, or real voltages for that matter. There are just variables (numbers) that represent such electrical properties, connected by means of an analytical equation known as Ohm's law. Can we call that a simulation or even a digital twin? Not so much.

Say we want to analyze how the circuit behaves if the ambient temperature drops by half. Our spreadsheet model cannot answer that, at least not yet. We must add to the spreadsheet the equation that relates resistance variation to temperature. After adding that, now we want to benchmark how the real circuit in the lab and our model behave as we arbitrarily vary the temperature. We quickly see a problem: the spreadsheet reacts instantaneously to the temperature variation (which is now an input), whereas the real circuit does not. Why? Well, we have not added any concept of "time" to the model; it just instantaneously calculates resistance, voltages, and currents as we change the inputs. If we want to add time to it, we have to surely drop the spreadsheet and somehow code so that our model can make the variables of interest evolve with time. This would lead us to a set of ordinary differential equations, which also means we will have to find a way of integrating such equations to obtain the output variables. Say we do that, and now that we have a model that is aware of time: note that the notion of "time" in a simulation model is purely a construction; a discretized construction. Whereas the real circuit on the table can only evolve in wall-clock time (it cannot evolve faster than what the environment and its own time constants dictate), a simulated model can evolve as fast as the computing resources allow. In the real circuit on the table, time elapses continuously, but in our computer simulation model, it elapses in discrete steps. During a simulated timestep, the computer model must calculate all the relevant state variables and guarantee the results are ready before the next step comes. What happens between two subsequent steps in the simulation? Nothing; all simulation variables remain frozen with the last computed values until the next tick comes. Simulated time can also run backward.

Back to our circuit, imagine we wanted to measure how fast the voltage drop on the resistor settles to its final value while turning the switch on and off. We measure the real circuit, then we check our model, and we see a difference. Why? Well, our model has so far considered that the wires connecting the different elements are ideal (no resistance, no inductance, no capacitance). We purposely decided that wires just "do not exist" in our model, and things were linked sort of magically; now we are paying the price for this simplification. Turns out, wires do seem to exist, and they have physical properties that affect the behavior of the circuit. We have been able to ignore this only because our focus was somewhere else. As soon as our analysis puts the focus there, all of a sudden, our model's coarseness becomes very evident.

System analysis and modeling largely rely on this opportunistic thinking: if conditions allow, we can ignore lower-order effects until we cannot. It is a modeling challenge to keep this under control as the need for fidelity evolves. But it is also a challenge (and perhaps the most difficult one) to understand where the line lies between adding useful vs. pointless realism to our models. From a computing perspective, unnecessary features of a model consume valuable resources and must be left out.

Finally, let's say we want to create a full-digital simulation of our electric circuit. We want to make sure the model will behave, in every single aspect, the same as the circuit on the lab table. This means that we would have to simulate the battery chemistry, the wires' distributed inductance and capacitance per unit of infinitesimal length, the switch mechanics, mass and stiffness, the resistor internal composition, and solve for all the partial differential equations involved. This could snowball into a feature-adding spree until we find ourselves modeling down to the quantum mechanics of the most elementary particles known. We have thus reached the point Paul Valéry was trying to warn us about: the model has become unusable.

This leads to some conclusions:

- As we increase agency in our models, we are required to understand beforehand how the new agents will behave. In the circuit example, to be able to model the behavior of the circuit more realistically, we need to know what effects are missing before adding them. For example, we must know Joule's law exists before adding that feature to the circuit model.
- As we increase agency in our models and simulations, every new agent has its own state vector, which is a set of numerical variables that describe its past, current, and future as simulated time advances, and a computing resource must keep evolving accordingly. The past might be kept in non-volatile storage for analysis and replay, either forward or backward. Then, a "global supervisor" of sorts must ensure the cohesion between all the agents; i.e., enforce the rules on how the agents can or cannot interact. Even if we try to abandon the "global supervisor" idea and ideally assume full parallel computing, where we assign to each agent its own dedicated computing resource, it still does not resolve the problem of needing something else to decide how those agents are supposed to behave with each other. In our physical reality, physical laws are pervasive, and they dictate how all particles in the known Universe are supposed to behave when they mingle. In simulated realities, those "laws" must be constructed and coded as well. Nobody presently knows how to reproduce General Relativity and particle physics from an algorithm running on some sort of machine; we can only *approximate* the laws that we *know* with a computer simulation. Attempts to algorithmically reproduce natural laws can quickly face incompatibilities related to the symmetries of theories of special and general relativity, as well as "spooky actions at a distance".
- Will we ever be able to manage to model things exactly as they are and create true digital twins that can fully mimic a physical system? A perfect model should generate the same information as the real system, given the same stimuli. But as we said before, even two copies of a physical system will not generate the exact same information given the same stimuli, nor will they show the exact same

behavior, due to the compound nature of such systems. Every copy of a physical system is, to a statistical level, different. As functional complexity grows, the dispersion between copies of a physical system will also grow. This means that an eventual perfect digital twin would need to either be perfectly tuned for each specific instance of a physical system or accept to compute information with a tolerable error.
- Could an operator be unable to tell if they are operating a digital twin or the real thing? It would require insane levels of immersion to get out of the digital twins' deep "uncanny valley" they're in. Think of this experiment: what would it take for a pilot to be unable to say if they are sitting on a flight simulator or a real aircraft? How far are we from that?

4.3 Hype Me Like One of Your French Girls

When discussing fads and fashions, we just cannot leave the topic without discussing hype. Hype is excessive or exaggerated publicity and excitement around something, often creating inflated expectations that may not match reality, and it's a remarkable creation. A well-oiled, global, distributed machine that brings bread to the table of millions of families worldwide. The beauty of the hype engine is in its elegant simplicity: nothing needs to be manufactured, only excitement, materials, and content. No factories, no inventory, no bolts or washers. PowerPoints, brochures, and loose terminology. And it generates billions year over year.

To start the fire, an initial spark is needed, typically from one of the big business consultancy behemoths, identifying a new technological trend. Said trend needs to fulfill a few attributes:

- It must be early and vague enough to avoid making the lack of traction and real metrics too visible. If people can actually measure or size it in any way, you're too late.
- It must be framed as cutting-edge and beneficial for our lives (as opposed to any technology easily related to war, death, or destruction; those are harder to hype, though admittedly not impossible).
- It must be capable of hosting an obscure breadth of terms and complex inner workings that are difficult to grasp for the layperson. Terminology complexity is a good camouflage; if they understood it, they might ask uncomfortable questions.

If these conditions are met, eureka: the "hype flywheel" kickstarts and starts manufacturing reports and white papers left and right, showcasing the undeniable potential of this new technology and how it will change our lives forever. Examples of what the hype machinery has been producing in recent years include (note that many overlap in time):

- Experience Curve (1970s)
- Semantic Web (late 1990s)
- Big Data (2007–2011)

- Internet-of-Things and "Industry 4.0" (2011–2015)
- Blockchain (2015–2018)
- Quantum Computing (2018–present)
- The Metaverse (2020–present)
- Artificial Intelligence (2019–present, although AI is a recurring hype they have been recycling for the last 25 years or so).

Clever hype makers never dive into any of these in isolation but connect them for a bigger yield. "AI-powered blockchain solutions for the Industrial Metaverse", or one I saw recently, which is brilliantly compact: "Edge AI to transform the Industry 4.0". Amazing.

The fundamental parts of the hype mechanism are as follows:

- A good number of articles must be published in renowned quarterly business reviews. With the wheel spinning at full speed, news starts going around. The smaller fish—the layperson who subscribes to those outlets—start consuming the smoke.
- Spurious reports filled with doctored market sizes and growth rates proliferate. The numbers are too good to look away. Billions, trillions, of dollars by a certain year which is always at least a decade away (and shifts accordingly).
- Early entrepreneurs spot the edge and start pitching. No product, no idea if the technology is feasible, but the vision is big. Revolutionize, democratize, disrupt! It's cheap to machine gun verbs, at least cheaper than paying engineers to do real work.
- Venture capitalists, advised by dubious consultants or guided by their own capitalist smelling sense, meet the entrepreneurs. Hands are shaken, deals are closed. The tech is now getting funded since the big players are betting on it. Still no product, but that doesn't matter; it never mattered.

Companies ride this wave, and specialized media play an integral part in it, acting as the substrate, the vehicle on top of which the hype reaches massive audiences. All while holding absolutely no accountability; cashing in and staying safely aside if eventually there's a carnage down the road if the content published turned out to be pure snake oil and investors get the rug pulled under their feet.

Tech and business media managed to devise a fine-tuned system where companies pay for a piece in some popular, specialized publication announcing something along the latest hot fad. Or it can also be the announcement of a new facility of thousands of square meters to be built somewhere far to manage the insurmountable peak of demand the company is experiencing, so they can mass produce whatever they do, at the thousands per year. Always adding fancy 3D renders—and frequently forgetting to clarify they are renders—hoping the textures are right enough for an untrained eye to confuse them with something real. In some other cases, it might be a piece where two or more opaque organizations announce they are joining forces, with flashy "partnerships" of unspecified scope that squeeze all the buzzwords in, including things such as "Revolutionary AI-Powered Supply Chain Transformation". A recent trend is also to announce technological "breakthroughs" without posting any evidence at all, adding fake names of engineers in the text that are ghosts you can't seem to find in any social media or papers anywhere, adding "results" which are mere bulletized lists of too-good-to-be-true outcomes without further elaboration.

All this is nothing but the good old bluffing. Most engineers tend to despise bluffing; however, as much as they might dislike it, one must accept it as an unavoidable practice that lies at the very core of pretty much whatever we embark on as a society.

Bluffing is a big, hidden factor of production. It impacts and influences the world as a system, and it's part of our everyday life. Bluffing helps land jobs, gets companies funded, finds dating partners, gets presidents elected, and starts and ends wars.

But what is bluffing? It means deliberately twisting and decorating some facts for achieving somegain. The interesting thing about bluffing is that it may sound negative or unethical at first, but it can also sound extremely noble and even necessary, depending on the context.

Think of the ghost army made of rubber tanks used to fool the Germans during WWII. No one would call that unethical. Just as no one would consider it unethical to make yourself bigger and taller than you actually are while facing a black bear. When the bear is about to charge, you wouldn't think, "Mmm, this is unethical. This is not truly *me*". You'd be a snack in no time. So, if you want to spot a bluffer, look no further: check the mirror tomorrow morning. We all do it.

Bluffing finds its way into business life so naturally that we almost don't even think about it. In any "race" where contenders compete for a resource that only a few can get, the grounds for bluffing bloom in the absence of better competitive advantages. Chances are, we don't have the skills or charm our immediate competitors have. Then, we are left with one option in the toolbox: make ourselves appear shinier, bigger, smarter, better, cheaper, more attractive, more unique, more ingenious, more cutting-edge than others. More than we are in reality.

Is bluffing a recent trait of our culture, or something resulting from modernity or recent sociological shifts? Of course not. Bluffing is as old as time; Greek mythology has several tales about it.[3]

And what place do engineers take during a hype peak? When it's running hot and the bluffing skyrockets, engineers have the uncomfortable role of trying to bring substance to the otherwise smoky claims in press releases. Since hype may bring funds and fat budgets—at least for as long as the hype is relevant—If clever enough, engineers can cash out from the wave with good perks and salaries, so why not? All while always being extremely careful not to end up being too close with the occasional swindler and ruin your reputation in the process.

A world without bluffing is utopian. Fine, let's not be too idealistic and think of a world with zero bluffing, as long as we remember to "walk the talk" once reality eventually kicks in. In engineering, there is no harm in some strategic bluffing to fight the eventual "black bears" we face as we go about doing what we do. As long as we put our money where our mouths are. Otherwise, it's just fraud.

---

[3] For instance, see the tale of Prometheus, Dolos, and the sculpture of Veritas.

Chapter 5
The Velocity Fallacy

Abstract We have been taught the lie that good engineering takes time. That the only way is through endless audit, overdesign, overplanning, and risk avoidance. Reality is that this slowness is, most of the time, a result of overhead and a general lack of urgency. Traditional engineering approaches, especially in aerospace and other safety–critical domains, drag their feet as they happily underutilize humans as paper pushers and for taking care of slow, sequential tasks that machines could easily do and do better.

Keywords Iteration · Learning velocity · Organizational design · Rapid prototyping · Design cycles · Urgency

Traditional engineering approaches, particularly in aerospace and other high-stakes domains, normally operate on decade-long timelines. In those domains, people retire before they can get to see the result of their work. This legacy system rewards professional patience with job security and stability, a corporate model that is being incrementally obsoleted by engineering professionals who are not so willing anymore to see their career years inch painfully slow like a sloth on a hot summer day.

The slow pace of the legacy system does reflect real complexities and risks present (including public safety risks). However, and not sure when, slowness became a synonym of thoroughness, when such a correlation is, and has always been, rather spurious.

In traditional engineering, things are slow by definition, all while the brilliant human mind is blatantly misused as a slow, inefficient sequential machine. We humans are amazing at intricate, nuanced thinking processes, but we are slow at manual tasks like adding data to a spreadsheet or inputting commands in a dashboard. I have had the sad privilege to see humans in traditional engineering projects executing manual procedures of thousands of pages, one by one, in a monumental waste of gray matter and time. No wonder things take decades.

Recent developments across industries have redefined the old speed versus thoroughness relationship and ultimately shown that velocity and success are not as related as some graybeards wanted us to think. The world's most famous company

I. Chechile, *Engineering is Broken*, https://doi.org/10.1007/978-3-032-18169-5_5

building reusable rockets has compressed traditional development cycles by a landslide while achieving good reliability, including systems that carry humans to space stations, leaving many graybeards scratching their bald heads. A while ago, this same company revealed a rocket engine that looked so minimalistic that many veterans of the industry claimed, in absolute disbelief, that it had been photographed intentionally incomplete. Until they showed it running.

There is no magic here: organizations that focus ruthlessly on rapid iteration and learning from reality outperform those that prioritize endless planning and risk elimination through redundant analysis.[1]

Technical progress is fundamentally a function of iteration count multiplied by improvement per iteration. More cycles mean more learning and faster progress. Traditional approaches might spend years analyzing and modeling to build the perfect thing, while iterative companies build many imperfect ones rapidly, test them against the elements, acquire data, and rebuild. This way, bad designs reveal themselves quickly and cheaply. Good designs surface faster because they're tested against real conditions, not numerical models and simulations. Each cycle carries data that informs the next, creating a virtuous cycle that analysis alone cannot match. In fact, the brightest do this: they feed simulations—which are obviously still necessary—with real data collected during tests.

This all needs re-signifying our relationship with failure. When each iteration takes years and costs millions, failure becomes undesirable and damaging, also financially. The whole work becomes a quest for preventing failure. The cost of being wrong grows too high. When you make iteration cheap and fast, you make failure more affordable and learning an inevitable outcome. Reality will always surprise you in ways models can never predict.

Also, one of the most consistently missed points in the history of engineering is the organizational design behind the engineering craft: The machine that designs the machine.

Where traditional approaches love routing decisions through meandering thick layers of insecure management and across silos and empires, fast organizations lean on smaller groups to make decisions and implement solutions quickly. There is no speed possible if decision-making is so touchy-feely with hierarchy that everything must wait to be "escalated" for approval from the Gods Almighty. Engineers must be encouraged to try solutions and learn from results. The entire system must be architected to compress cycle time at every level.

I get it: such fluidity can be scary, and it's not for everybody. In a fast organization, if you ask an engineer about a design detail today, the answer might differ next month because the design has surely evolved. Traditional engineering interprets this as bad, no bueno, as instability, as poor planning. But constant evolution is the natural state of a system under design.

---

[1] In a beautiful irony, the competing spacecraft to carry humans to the International Space Station is made by a "traditional" company whose (delayed) design has shown flaws and quality issues left and right and is, at the time of writing, not certified for human spaceflight.

The cost of being slow is both schedule cost and the opportunity cost. Those years spent in analysis paralysis and planning accumulate costs in ways budget lines don't tend to capture. By the time a ten-year development cycle completes, requirements may have changed, technology may be outdated, and markets may have moved on. A380s are being scrapped as we speak; a marvelous piece of complex engineering that was ready later than when the market needed it. While you are building the perfect gizmo, your competitors are iterating rapidly and learning lessons you (or the committee you formed) are still thinking about.

Development speed doesn't intrinsically increase risk. Slowness doesn't eliminate it either. Snail-paced engineering processes still produce failures, and they're more expensive and harder to learn from.

Learning requires quick contact with reality. Models are useful but wrong, and expertise develops through iteration and not from transcendental contemplation.As Henry Petroski put it: as engineers, we are in a position to change the world, not just study it.

5.1 Do It Now

Except few exceptions,[2] many great engineering achievements that shaped our world share a quality that has become a rarity in the modern engineering practice: they were done with a crushing sense of urgency. When the stakes were big and the timeline was measured in months rather than years, engineers found a way to cut the crap, the endless planning, the committees, and the overthinking that now permeates so much of our technical work.

They built the atomic bomb in less than three years. They constructed cargo ships in weeks. And they accomplish all that with no issue-tracking boards or Agile manifestos.

Wartime engineering operates under a different set of rules because it has to. There is no time for perfection, no tolerance for organizational nonsense, no patience for pointless meetings. When German U-boats were sinking Allied shipping faster than it could be replaced, Henry Kaiser didn't start hiring scrum masters. He revolutionized shipbuilding by applying assembly line techniques to vessel construction, standardizing components, and training workers who had never set foot in a shipyard before. The result was ships that weren't exactly elegant or innovative in their design, but they floated, they carried cargo, and they were available when they were needed. If that isn't true engineering, I don't know what is.

---

[2] Notable examples of long-duration engineering projects include the Sagrada Família in Barcelona (1882-present, 142 + years), the Great Wall of China (constructed over approximately 2,000 years from the seventh century BC through 1644), Cologne Cathedral in Germany (1248–1880, 632 years with a 300-year pause), the Panama Canal (34 years across two attempts, 1880–1914), and the James Webb Space Telescope (approximately 25–30 years from initial concepts in the mid-1990s to its launch in 2021). These projects often outlasted their original architects and required generations to complete.

The Liberty ships were considered ugly by traditional shipbuilders, but they simply won the war. Ugly works. Perfection was a luxury the Allies couldn't afford.

The same principle governed the development of countless other wartime innovations. Engineers and scientists worked with an intensity that seems almost incomprehensible today, not exactly because they were more talented or better educated than their modern counterparts, but because delay was a literal matter of life or death. The feedback loop between idea and implementation was measured in days. If something didn't work, you found out immediately and tried something else. There was no time to get lost in endless analysis, no opportunity to let bureaucratic nonsense slow things down. The war imposed a discipline that modern engineering has forgotten, which is the lost art of finishing things with the straightest line possible from ideation to deployment. Not sure when, but at some point in history, we learned to zig-zag, and we forgot to go any other way.

This isn't to romanticize war or suggest that we have to be terminally ill or that something like an asteroid needs to be *en route* to destroy us all to get the right sense of urgency. Rather, it's to emphasize that we *can* work with urgency when we have to, or when we want to. Urgency forces natural prioritization and exposes what matters and what doesn't. It strips away the endless layers of process that accumulate, each one justified in isolation but collectively slowing everything to a stop. The modern engineering organization is drowning in pure drag: meetings, eternal audits, and snail-paced reviews. Each of these may have its place, but together they make timelines painfully elastic, where starting is hard, and finish lines move all the time to the right.

Engineering needs to embrace a "do it now" philosophy, and the change of mindset must start by remembering Parkinson's law: tasks always expand to fill the time we have, and that our default instinct to postpone, to wait for perfect conditions, is often rooted in insecurity disguised as perfectionism. We must re-learn the art of knowing what corners to cut, without crashing into a wall.

This doesn't necessarily mean shipping garbage. As we will discuss, there's a point in every project where additional planning produces diminishing returns, where the best way to get good feedback about a design is to build it and see what happens. Sometimes, the gap between a working prototype and a product is smaller than it seems. Action beats eternal contemplation every day. The wartime engineers knew this instinctively because they had no choice; for some of them, bombers were flying over their heads. We have a choice, which is precisely why we need to cultivate the same urgency they felt, minus the violence.

Consider how much time in a typical engineering organization is spent on activities that would be instantly recognized as wasteful if we operated under tight time constraints. Meetings with no purpose. Emails that could have been a Slack message. Slack messages that could have been just silent thoughts. Design documents that no one will ever read. Future-proofing for edge cases that will never materialize. Circular debates about naming conventions. All of this happens because we lack a shared sense of what we're racing against. In the absence of genuine urgency, we expand work to fill the void.

We might not be at war, and we might be healthy. No asteroid is hopefully heading our way. That's great. However, the reality is that we do face urgent problems. Environmental, organizational, personal.

We can recover engineering's sense of urgency if every individual engineer decides that they will no longer participate in pretend work. That moving ahead is better than asking for eternal permission from the Gods of Hierarchy. That shipping comes before writing a perfect design document.

Time is a non-renewable resource; every day spent polishing instead of shipping is a day we'll never get back.

Do the work now. Reply today. Write the code. Make LEDs blink. Delay is the enemy. Operate as if the stakes are high, because they are. The stakes are our finite lives, our limited time. Life is too damn short to sit in a meeting to hear someone repeat what they said in an email yesterday. To listen to someone lecturing you about what you know. With the right sense of urgency and a strong non-BS mindset, any market is grabbable and every competitor edible.

Chapter 6
Design Versus Engineering

Abstract People use the terms engineering, product, and design too loosely, when they mean different things. The design process is essentially a journey, a search process that leads us to identify needs (including our own needs), sketch possible solutions, test them, and refine them until we get something that works. It's undeniably creative, and it's not the exclusive territory of engineers. A product is anything you can sell to someone. Products are also not an exclusive concept stemming from the engineering domain. But then, what is engineering, and what do we get if we strip away the design and product aspects of it?

Keywords Design process · Engineering · Product · Fundamental principles · Design thinking

As soon as my professional career began to take shape, I noticed that my university education hadn't really explained much about the "formal" process through which technical objects are created. I had an idea based on common sense, but not much more about how one actually travels through what I now know is called the lifecycle of a system.

At university (I studied Electrical Engineering), I was force-fed physics, calculus, semiconductors, circuit theory, control systems, power, radiofrequency, and many other "fundamentals". And to be clear, I'm not criticizing that content; they're essential and useful. But somehow, in the syllabus, there wasn't anything about the process that turns a fuzzy idea into a well-functioning technical object.

With time, I realized this isn't something to blame engineering schools for. It's not really their domain. That mysterious, somewhat chaotic, sequential yet unavoidably iterative process is nothing but the old, messy design process.

In *To Engineer is Human*, Henry Petroski writes that "design is about getting from here to there" and that "the choices of design are ultimately like the choices of life".

Design is essentially a journey, a search process that leads us to identify needs (including our own needs), sketch possible solutions, test them, and refine them until we get something that works. It's undeniably creative. And here's a key point: it's not the exclusive territory of engineers. Design stands on its own, and as such, it

I. Chechile, *Engineering is Broken*, https://doi.org/10.1007/978-3-032-18169-5_6

can be applied to pretty much anything. Clothing. Furniture. Public policy. Pastries. Business models. The list goes on.

In truth, practically anything can be both the object and subject of design. Design is a deeply human skill. A 3-year-old toddler designs when they build a garage out of Legos for their cars. A prisoner designs when they quietly sketch an escape plan. Of course, the design process has to be adapted to the object under design. Designing a rocket is not the same as designing an escape plan for Alcatraz.

The design–engineering duality gets worse when the concept of *product* enters the scene. Product as in: "something designed and developed for commercial purposes". Engineering and product development are related, sure, but they're not the same thing. A space shuttle is a marvel of engineering. However, it is not a product.

So what is engineering if we strip away the design process and the product side of it? At first glance, it feels like not much is left. But on a second reading, what remains is precisely the essence of it.

Engineering is the act of using fundamental principles combined with a good dose of heuristics to change the context in a way that something previously not possible becomes possible. But careful: this isn't necessarily about the platitude of "making the impossible possible". Everything engineering has ever achieved was always technically very possible. Otherwise, it would never have been achieved. What was missing was the ingenuity to make it happen while standing on the shoulders of science and lessons learned, dragging into the light what was hiding in the shadows.

For example,

- We couldn't call or message friends or loved ones until we could.
- We couldn't observe our planet "from the outside" until we could.
- No human could lift a millimeter off the ground without falling back down in a predictable parabola, until the stubborn Wright brothers made it happen.
- We couldn't regulate the heart's electrical rhythm with man-made machines until we could
- We couldn't reuse the first stage of rockets until we could
- We couldn't drive or take a train from France to the UK, or from Denmark and Sweden, until we could
- We couldn't integrate transistors on a single piece of silicon until we could
- We couldn't use the energy of nuclear reactions for power generation

The list could go on forever. Once engineering breaks the mold and shifts the context to make something possible, it's irreversible. There's no going back. Once it has shown it can be done, the game changes forever, and the quest becomes about improving it and making it more efficient. Soon after, it'll be productized. Commercialized. Or, alas, militarized.

Once engineering has figured things out, what the rest of us mortals mostly do is follow the recipe. And that, strictly speaking, is *not* engineering. That's just tracing the path.

A modern airliner didn't make flight possible.[1] The newest GPU didn't make it possible to pack complex circuits into a tiny slab of silicon. That doesn't make them any less impressive; they're spectacular objects. Are they better? Yes. Do they outperform their ancestors by orders of magnitude? Absolutely. But at their core, they're following the recipe. They're condemned to be unoriginal because, at the end of the day, they're reusing a concept or architecture someone else unearthed through ingenuity, experimentation, and trial and error. They're covers. Remixes. Not the original song. I know it may sound a bit reckless to say these technical wonders aren't "pure" engineering, but I'll die on that hill: at the macro level, they don't shift the context irreversibly.

I admit it; this idea of "real" engineering is quite close to invention, though it doesn't require any *Edisons* or *Nikola Teslas* fighting over which electrical current is better. You can break the mold through the sheer determination of ordinary people. Engineering isn't about inventing for invention's sake. It's about creating something that lets us do what we couldn't do before and do it in a reliable, safe, and repeatable way, often using existing, known concepts in novel ways.

[1] Let's stop the hyperbole here and be fair with modern aircraft; they are undeniably a symbol of great engineering. For instance, for the substantial use of composite materials, overall electrification, and on-board software.

Chapter 7
The Engineer Versus the Engineering Manager

Abstract The engineering practice lives at the intersection of the concrete and the abstract. While engineers solve tangible, well-defined problems, engineering managers deal with the intangible realm of strategies, priorities, and organizational affairs. This chapter explores the fundamental difference between these two dimensions of engineering work: the linear, first-in-first-out logic of technical problem solving versus the messy, interdependent, and evolving "networks of problems" that managers accumulate, as they juggle with networked collections of problems, often without any tracking or closure. The result is a constant fight between those who design and build systems and those who manage them.

Keywords Engineering management · Systems thinking · Organizational behavior · Engineer–manager dynamics

The engineer who routes a circuit board or codes a software driver is doing a rather concrete task. The results of their work are quite tangible (whether you can touch them or not), and those problems are, in general, well-defined, although not necessarily easy to solve.

In contrast, an engineering manager who discusses plans and strategies does not enjoy such concreteness. In managerial roles, the fuzziness of the topics makes it rather hard to jump into action, as an engineer can jump into writing a line of code. The path for the manager is not as tangible.

In both cases—in the concrete life of the engineer and the abstract life of the engineering manager—they deal with problems. A problem, despite the negative connotation in its name, is not always necessarily a bad thing; it is simply something that must be solved, sorted out.

For solving concrete problems, the engineer applies a first-in-first-out approach: the first problem needs to be solved before the next problem can be tackled. Rarely will a design engineer let too many problems pile up before feeling their life has just gone out of control.

Engineering managers behave rather differently. Because of the fuzzy nature of the things they deal with, it is common for them to collect problems as they go.

I. Chechile, *Engineering is Broken*, https://doi.org/10.1007/978-3-032-18169-5_7

What is more, managers collect "networks of problems" that are related one way or another.

Russell Ackoff stated ages ago that managers are never confronted with problems that are independent of each other, but with situations that consist of complex systems of evolving problems that interact with each other. Ackoff calls those networks of problems *messes*. Problems are to messes what atoms are to tables. Managers do not solve problems; they manage messes.

So, engineering managers go around with their virtual baskets collecting problems and knitting messes, and they do not function first-in-first-out like the engineer, but more like in a last-in-first-out way. They pile them up and treat them the same way we pick socks from a drawer. Only the last, freshest ones might get attention for as long as there isn't a new problem coming in to push the current problem down the pile. They can't keep too many problems in their minds, so they only pay attention to the 3 or 4 most pressing ones. The rest? Down the drawer. Additionally, managers are simplification machines; whatever problem you bring to them, they will never retain the details and complexity your problem carries. They will strip all that and keep a very simplified, borderline childish version of what you initially conveyed. Note this is a built-in feature when managing messes, and not a critique; managing networks of problems requires simplifying the nodes of the network. It's just important to remember this when you feel your manager is only addressing about 5% of what you brought to them.

Once, I was struggling to find the proper organizational structure for an engineering team I was leading, so I brought the problem to my boss. He took it, and we didn't speak for a couple of days. After a while, I kindly reminded him about the issue, and he said, "Ah, yeah, I've been thinking about it, and made a diagram for you!", and then he proceeded to hand me a folded A4 sheet of paper. When I opened it, it was three boxes with some random acronyms I hadn't seen before, linked by some shaky lines with arrows. He said something like "What do you think?". Well, my 7-year-old kid could have drawn it better and could have been closer to solving my organizational design issues compared to that rudimentary piece of...art? Oh yes, before you ask: I gave him the possibility to explain the drawing, which didn't make any sense either.

Needless to say, when bringing problems to managers, you rarely get any "tracking" like software people use for their bugs; no ticket number, no schedule for completion, or status. All you get is a promise of a solution that will largely depend on their capacity to remember it as the next thing comes their way. Chances are they will forget.

It is in this battle between the concrete nature of tackling technical tasks the engineer deals with and the abstract nature of managing them that the manager deals with that a lot of energy is burned.

Far from being a defense to disorganized managers who don't care enough to give closure to what you bring to them, this only aims to highlight that the difference in the way engineers and engineering managers work is perhaps the biggest source of problems between the two, and a source of headaches and frustration. As an engineer, you expect your manager to treat the problem you bring up to them the same way you deal with your technical problems. As an engineering manager, you expect the engineers reporting to you to have the ability to understand abstract things skillfully and help you manage *messes* that only keep growing. You two work on different dimensions and with different prioritization algorithms. Somehow, you need to learn to meet somewhere in the middle.

Chapter 8
Surviving Cringe

Abstract Many engineering managers are unaware that one of the main traits of being a manager is navigating the ultra-thin line between being inspiring and being unforgivably cringeworthy. What is that managerial cringe after all, and why does it hit so hard? Managerial cringe hits you when the fake, artificial part of a message from a higher-up rings louder than what's being said. It can be an artificially epic quote or the excessive use of pseudo-heroic figures of speech, including lions, wolves, and warriors. A manufactured message thrown in Slack to say something the world would have totally survived from if gone unsaid. It can be a quote from Sun Tzu, or a meme used the wrong way. In the engineering workplace, cringe comes in many different forms and flavors.

Keywords Communication · Cringe · Legitimacy · Leadership · Credibility · Communication

If you search around in journals of management for the word *cringeworthy*, you get zero results. Nada. The academic machine, with all its peer-review etiquette, has produced entire literatures on leadership theory, psychological safety, sensemaking, agile transformation, and "servant leadership", whatever that is, yet not a single paper on that familiar full-body recoil you feel when a manager says or does something so ridiculous that your soul tries to eject from your body.

I bet you have felt it. It's that strong spasm your body does when you hear a boss say something so awkward, so unnatural, that you want to crawl under a table. It can be a vapid, artificially epic quote or the excessive use of pseudo-heroic figures of speech, including lions, wolves, and warriors. A manufactured message thrown in Slack to say something the world would have totally survived from if gone unsaid. It can be a quote from Sun Tzu, or a meme used the wrong way. It can be a grammatical abomination in some company-wide presentation. Can be the systematic use of buzzwords: *data-driven, revolutionary, next-generation, actionable, going forward, synergies, low-hanging fruit*. It can come from witnessing someone trying to explain something technical while extremely out of their depth. It can be a poorly crafted email whose idea is ugly and tone-deaf. It can be a forced team outing activity where you find

I. Chechile, *Engineering is Broken*, https://doi.org/10.1007/978-3-032-18169-5_8

yourself camping out with your boss while he won't stop talking about himself around a bonfire, and you wish a bear would jump out from the trees and eat you just to grant you some mercy. It can be your manager's shameless adulation of someone higher up the hierarchical ladder.

Cringe comes in many different form factors and flavors. Tastes horrible every single time.

Many managers are unaware that one of the main traits of being a manager is navigating the ultra-thin line between being inspiring and being unforgivably cringeworthy.

What is that managerial cringe after all, and why does it hit so hard? Since we cannot rely on research because there isn't much, let's do some qualitative analysis here of what cringeworthy management is. Besides the typos (which we can blame on their lack of proofreading or even lack of care to pay attention to the autocorrector), managerial cringe hits you when the fake, artificial part of a message rings louder than what's actually being said. And it rings loud because of legitimacy; the lack thereof. When the legitimacy of the source is dubious, when he or she speaks from a position that is not exposed to the elements as the ones they are addressing and asking sacrifices from, the cringe factor soars because their epic narrative sounds with the echo of a massive, empty room.

Legitimacy helps mitigate the cringe but does not completely nullify it. As legitimate as you could manage to be as an engineering manager, you can still be cringe, so it is always good to avoid the manufactured grandeur, the corny quotes, the platitudes, the bellicose analogies, the vapid mottoes. Just…just don't.

Engineers survive the cringe with some selective internal filtering. They nod at the Sun Tzu quote. They endure the bonfire monologue. They swallow the buzzwords. They try hard to translate all that into real work.

Chapter 9
The Pretend Work Era

Abstract As we speak, backlogs in issue-ticketing systems are inflating at unmanageable rates. Parking inspectors are not the only ones ticketing for a living. In the engineering workplace, certain people attend meetings and return inspired to their desks, machine-gunning a stream of tickets into your inbox that will distract you from what really needs to be done. The issue-tracking ticket has become the currency of *busyness,* or the obscure art of looking busy. Meanwhile, real work is never captured in those digital sticky notes; work that still must be done because that's what customers pay for.

Keywords Issue-tracking systems · Engineering ticket · Productivity · Workflows · Pretend work · Productivity · Busyness

Just like a photo of a sandwich won't stop you from starving, a handful of tickets in an issue-tracking application doesn't build you a product. It's shocking how we seem to have forgotten that.

In these trying times, teams spend their days shuffling poorly-crafted tickets around "Kanban boards", patting themselves on the shoulder and congratulating each other for clicking buttons and moving virtual rectangles from "To Do" to "Done", creating an illusion of action. You see people celebrating charts out of those tools as if these were actual indicators of progress. There are board *connoisseurs* who are purposely hired for fine-tuning workflows, filters, automations, and conditional color rules. Some organizations hire entire teams whose sole function is to administrate this rapidly expanding paraphernalia of virtual work. Virtual as in ineffective and ethereal.

As we speak, backlogs are being inflated beyond human comprehension. Full of tickets titled with vague imperatives like "Refactor logic" or "Add logging feature" or this one I saw once: "Define architecture" (do you really need a ticket for such a fundamental task[1]?). Tasks so open-ended they could mean anything or nothing,

[1] Once, I also saw someone having a full weekly schedule made with sticky notes on a wall. His typical day would include a sticky note saying"Lunch". I have always wondered if this person could have starved to death, had the sticky note not been there to remind him to eat.

I. Chechile, *Engineering is Broken*, https://doi.org/10.1007/978-3-032-18169-5_9

depending on who's reading them. Certain people don't build things anymore, they ticket for a living; they're corporate parking inspectors. They attend meetings, return inspired to their desks, and machine-gun a stream of tickets into your inbox that will distract you from what really must be done, because you can't reply to a ticket and do actual work at the same time. It's either one or the other.

The issue-tracking ticket, in this twisted reality, has become the currency of *busy-ness,* or the obscure art of appearing busy. Writing a ticket feels like you belong to the pack. Commenting on it feels like you are contributing to global GDP. Creating or modifying a workflow is, wow, you feel like God. But it is all just digital bureaucracy disguised as sophistication. Meanwhile, work—the messy, analog, deeply human process of thinking through problems, orchestrating code, debugging, testing, arguing, doing tradeoffs—is never captured in those boxes. Too human, I guess.

At any rate, most tickets meet the same tragic fate: they rot to death. Every engineer knows the smell of a rotting ticket. They all begin bombastically: nicely worded, tagged, prioritized, and assigned to someone without even asking. But then, context shifts. Priorities mutate. Shit happens. Weeks pass. The ticket's relevance fades until it becomes a relic of a past that is never coming back; a digital trace of something that was never delivered or needed in the first place.

In the meantime, work had to be done in the analog world because that's what customers pay for, because customers don't pay for Kanban boards or "burndown charts". Real work that was never captured by any of those tickets, to the silent sigh of the issue-tracking lovers, who would love engineering to be about the sticky notes they shoot, while they intimately know that they don't drive real work at the end of the day.

Chapter 10
Prototypes and Products

Abstract The journey from prototype to product describe sold challenges in the engineering practice. Rapid iteration and naiveté may prove technical feasibility but may also reveal market unfitness. Engineering organizations face a dilemma: deploy minimally viable gizmos that barely work and improve through real-world use or polish until perfection before launch. How and when should "skunkworks engineering" projects get out of the dungeons and hit the shelves? How does a company go back to the R&D table after a market winner?

Keywords Prototyping · Iteration · Minimum viable product · Productization · Time to market · Skunkworks engineering · Product development · Commercialization · Value chain · Early-stage engineering

Building a human-powered airplane was the challenge set forth by Henry Kremer in 1959. For 18 years, nobody could do it. But within six months of trying, Paul MacCready built and flew his Gossamer Condor. The difference was in his approach: while others needed a year's worth of effort for each test flight, he created a plane that he could fly, fix, and fly again in a matter of hours.

The story of how MacCready won the prize is interesting. While other teams had more people, time, and resources, which enabled them to make sophisticated aircraft, that didn't get them close to winning the prize. McCready noted that his inexperience and lack of resources were, in fact, a strength. While other teams relied on highly specialized experts applying conventional aircraft design methods, this approach deliberately avoided established disciplinary frameworks. Lacking formal background in wing structures became an advantage: by starting from first principles and ignoring how aircraft "were supposed" to be built, it was possible to rethink the problem entirely. Drawing only on basic physical reasoning and informal experience with lightweight flying machines, the result was an unconventional design that achieved an extreme span-to-weight ratio experts would have discarded in their analyses. Deeper specialization obscured simpler solutions, as prior knowledge constrained the space of feasible approaches.

I. Chechile, *Engineering is Broken*, https://doi.org/10.1007/978-3-032-18169-5_10

A somewhat simple read of the story might lead us to the conclusion that the Gossamer Condor is yet another proof that with naiveté, fast iterations, and incremental design, everything is possible. And that interpretation would not be completely wrong, although there's more than what meets the eye. However, why aren't we all pedaling through the air to work every morning if human-powered flight were such an easy thing? Why aren't there hundreds of Gossamer Condor-like human-powered tiny airplanes parked outside railway stations in Amsterdam, Brussels, or Buenos Aires? The Condor was just a rough prototype. It was unsafe, rudimentary, and fragile, all inescapable—and tolerable—traits of prototypes. Prototypes are made to prove a point and be as quick and cheap as possible in doing so. Safety, compliance, and ergonomics are never part of the thinking process during prototyping; all things that are essential in production.

In time, MacCready and his team improved the idea and designed the Gossamer Albatross, with which they aimed to cross the English Channel.[1] The Albatross was an undeniable evolution of the Condor, but still, it was nothing remotely near production level. Acknowledging the difficulty of human-powered flight, MacCready would eventually shift into solar-powered electric aircraft with his subsequent models like the Penguin and then the Solar Challenger, early projects of what would then become AeroVironment, his aerospace company. Neither the Condor nor the Albatross went commercial in the end. But that's the idea behind prototypes: they might provide the uncomfortable confirmation that there is no business for them.

Too many times we see prototypes dressed as products, but one thing is to come up with a rough strawman concept, and an absolutely different story is to productize that into a product. Products are things people pay for, and if money changes hands, you'd better take care of everything.

The Gossamer Albatross involuntarily made a beautiful point when it struggled crossing the English Channel. Prototyping can help you to prove a point. But hacky gizmos will not provide you with enough "thrust" to cross the prototype-product gap. Let's talk more about that gap in the sections below.

10.1 You're the Real MVP (Minimum Viable Product)

Now, imagine that your prototype does seem to have a commercial edge. When is it a good time to release it?

There is an enduring cold war confronting two opposing ideologies in the engineering world between those who embrace an incremental approach versus those who intend to get it "right the first time". We spoke about this when we addressed the Velocity Fallacy. Now we are more into when than how.

---

[1] On June 12, 1979, amateur cyclist and pilot Bryan Allen successfully flew the Albatross human-powered aircraft across the English Channel, despite radio failure, dehydration, and deteriorating flight conditions, landing on a beach in France at 30 km/h from an altitude of 1.5 m.

In other words, the criterion to define *when* something can be released to an audience. And here it is important to state the obvious: all engineering approaches are iterative and incremental at their very core. No one does it in a linear, truly unidirectional way; we all go back to previous steps to make changes, rerun part of the work, and modify stuff around. The key difference lies in where to draw the line when deciding to launch something versus keeping it in the "hangar" until we have smoothed the last rough edge, and the thing is immaculate and ready to hit the users or customers.

The obligatory platitude says that *better is the enemy of good enough.* There is nothing wrong with preferring better over good enough. Of course, as long as you can afford better. Otherwise, you have no choice but to accept *good enough.* But here's some good news: the sole act of deploying *good enough* stuff will pay off, because the sole act of being out there in the elements will exert the necessary forces to incrementally improve it. Those insisting on deploying the most perfect artifact ever conceived by humankind while having little time and shallow pockets are plain suicidal.

However, you can't release something that just doesn't work. Therefore, there is a minimum viability threshold that must be reached before something is ready to hit the road and be operated, used, or consumed. There might be safety and regulatory constraints to prevent us from launching something in the wild before certain features are put in place for it to be safe. However, if we are not dealing with anything mission-critical, once we reach said minimal viability, can't we just roll it out and perfect it further from there, with users and customers in the loop? Here's when the ideological dichotomy kicks in; some say hell yeah, some others say hell no.

Minimum viability is always an end-to-end thing. Forget for a second about engineering: imagine you want to sell pizzas. When do you reach your 'minimal viability' as a pizza guy? Right when the first pizza rolls out of the oven? Not really. If you want to make a living out of selling pizza in a professional capacity, you will only reach minimum viability level once the oven is in place, when you have a place with city approval, a way of billing, a way of packaging, a way of taking orders either from a counter, or from an app, tables and chairs, cutlery, glasses, neon signs, printed menus, beverages, a fridge, toilets, etcetera. True minimum viability is only reached once the entire value chain reaches viability level. It makes no sense to have one element in the chain at "perfection" level where all the rest is hanging from threads. Surprisingly, many engineers tend to forget that most, if not all, objects they engineer are part of a long value chain that must be stable for the thing to put food on the table. Engineers miss this point consistently.

Bringing a complex prototype to the market is a challenging task, and perhaps the most difficult hurdle in that journey is securing the very first customer. While conventional wisdom suggests that "early bird gets the worm", in practice, early customers are exposed to high risk and uncertainty. Selling that "first unit" requires more than good technical specs; it demands a good understanding of human psychology, economics, and risk management. It's often the second sale—not the first—that marks the true beginning of commercial viability.

Potential customers see adopting an early complex product as a high-stakes transaction. The first customer effectively places a bet on an unproven system. If it works, they may gain a competitive edge; if it fails, well they absorb not just the financial loss but also the reputational damage and opportunity cost.

This dynamic creates an intrinsic disadvantage for the first sale ever. While early adopters may receive discounts or favorable terms, these benefits often do not outweigh the risks. No one wants to be the guinea pig, especially for mission-critical systems.

This problem gets worse when the organization behind the product is also early-stage. When both the product and the company are unproven, the perceived risk skyrockets. In this case, customers aren't just betting on the functionality of the product; on top of all that, they're betting on the survival, maturity, and support capacity of the vendor behind the product. In contrast, if a proven company with a strong track record of delivering complex solutions releases a new product they haven't done before, that first sale becomes substantially easier, simply because it's not their first rodeo and they already have the structure needed for venturing into new things; reputation acts as a masking factor of product maturity.

The reality is that, in complex technical systems, the first customer participates, knowingly or not, in the product development. It's only after this trial by fire that the product matures into something manufacturable and scalable.

10.2 Productize This

Years ago, as a welcome gift on my first day at a new job, my new boss threw a huge blob of code at me. The code belonged to a tool that was originally conceived purely for internal use, and my boss asked me to turn it into a product that, as I learned in that very moment, already had waiting customers and whose delivery would unlock the project's payment schedule. Lovely pressure for a first day at the job.

Additionally, right after tossing me the code, the boss asked me to "close it", i.e., to protect the intellectual property, but not too much because the customer wanted to build their own logic on top of it (basically: create a software development kit, or SDK). The bosses also wanted proper documentation, a user manual, etc.

With no time to protest—being the new guy—I embarked on one of the strangest tasks one can get: turning into a product something that was not meant to be a product. The wizards of corporate vocabulary call this *productization*.

When you're tasked with "productizing" something, you quickly discover the enormous chasm between something built for internal consumption and an actual product.

When developing a product, several critical considerations must be addressed from the outset.

First, establish a defined scope that clearly outlines what the product should and shouldn't include, along with a price that reflects thorough cost analysis rather than arbitrary figures or wishful thinking.

Many products require a decent post-sales plan that goes beyond simply selling. This includes defining what happens after delivery, addressing client return questions, establishing who handles phone support, and creating warranties and documentation that outline the rules for product failures, recalls, repairs, and returns with clear expiration dates and procedures for when customers need to stop calling.

Needless to say, documentation is essential, covering API guides, style guides, brand control documents, and similar materials. Quality assurance must incorporate basic safety considerations to prevent customer injury or harm, as negative publicity can severely damage the product's reputation.

Thought must be given to shipment, integration, and deployment, particularly if the product will be part of a larger architecture managed by the client or a third party, ensuring seamless integration. If the product contains anything hazardous (batteries, chemicals, explosives), there might be strict regulations which might take months to comply with.

Licensing and third-party dependencies require careful analysis; you need to determine whether you can legally sell something that relies on open-source library X under license type Y, and understand the implications if a client resells the product to Pyongyang or other restricted territories.

The product should provide at least a minimal but not user-unfriendly experience, offering some usability and intuitiveness. A configuration mechanism is necessary, but it shouldn't be merely a pile of cryptic XML files edited by hand in Vim. At least, it should provide clear instructions on how to exit Vim.

Finally, give the product a name that is unpretentious, easy to pronounce, and doesn't sound like genitalia in some other language.

10.3 From Skunkworks to Production, and Back

Any company that has survived for a decent number of years with a reasonable product-market fit eventually finds itself getting really good at squeezing what already sells while nervously thinking of what comes next. For survival, you have to keep one eye focused on the engine room of reliable products and another on the moving sands of emerging technologies, new market entrants, consumer behavior shifts, and new markets.

This can be stressful. Because the mindset, pace, and risks involved in running a stable product line are completely different from those required to explore the unknown. Yet both have to coexist. Exploiting existing strengths and exploring radical new opportunities needs handling and is not as simple as it sounds. In fact, it doesn't even sound simple. It sounds complicated, because it is complicated.

This is where skunkworks engineering makes a glorious entrance: small, semi-autonomous teams set up to chase ideas that the rest of the organization couldn't chase because they would take a century. Skunkworks are designed to be fast, flexible, and a little weird if you want. They're supposed to move away from the gravitational pull of the core business because if they don't, they'll get stuck in red tape instead of creating the new stuff. And everybody will quit because those cut for working in fluid R&D are not made for all the corporate nonsense.

However, skunkworks engineering can be a dangerous game, too; first, because it needs money to keep going, results are not immediate, and failure is common. And also, because organizations often don't think hard enough about what happens if the wild idea actually works. What if the prototype is so compelling that it now suddenly needs a roadmap?

The divide between secretive demo and production is where certain resentment can build up: the product team sees the skunkworks folks as disheveled hippies; the skunkworks team sees the product organization as neophytes who have no idea what they're talking about.

Researchers Charles O'Reilly and Michael Tushman studied this in detail and came to a practical conclusion: the companies that managed it best were what they called *ambidextrous organizations*. These companies kept their innovation units structurally separate from the core business. Different pace, different objectives, even somewhat different culture, but they tightly integrated them at the senior leadership level.

There's a proper timing for it, and many get it wrong. Some may start worrying about the future too early. Some may wait too long to innovate while polishing the cash cow forever. Some might push innovation too far, plainly ignore it, or merge it too early. In all these scenarios, the prototype can't cross *the gap*. However, when leaders get their act together—allocating capital, resources, and managing incentives and expectations—then things move.

The organizations that play the skunkworks/production duality game well tend to build bridges early, with product-minded thinkers embedded in rapid innovation teams from day one. They do not throw a prototype over the wall but start the transition while the idea is still evolving. And they embrace organizational duality, where separation is healthy. Additionally, they frame the quest in a way that it's clear they're not building an eternal prototype but building the future products of the company.

Every company that survives long enough eventually walks this tightrope, but many lose balance and fall. The trick is to keep adjusting. You want your skunkworks to be weird. You want product teams to be rigorous. What you don't want is to pretend to be both. Send red tape lovers to try to tame a skunkworks team, and you will get pissed people from both ends. Innovation is useless unless it ships. Advanced projects are pointless if they will always be in the shadows. They need to cross *the gap*, become a product, and earn their place on the website.

10.4 Sticky Designs

We just talked about companies that can't easily let go of their market winners and struggle to work on what's next. Well, here's a fact:

A design that works sticks around.

The problem appears when a design is increasingly obsoleted by the boundary conditions, but it lingers more than what common sense would deem reasonable, mainly because the effort to start a clean-sheet design is perceived as higher than putting a bit more lipstick on the familiar pig. Until, of course, reality explodes in everyone's face. Let's see an example.

Everybody is criticizing these days one of the major aerospace companies in the world. We know who it is, so no need to name names. I am not going to delve into the latest issues the company has faced; I will leave that to all the knowledgeable armchair experts who are flooding our feeds with articles about the "downfall" of this company, which I do not think it is the case as it is "too big to fail", but they surely have homework to do if they want to fix their engineering and fundamentally their credibility.

Aerospace hasn't been immune to the stupidity that has encircled the engineering practice, as it has reached many other engineering organizations of all sizes and colors across the globe. These companies have it more difficult because the criticality level of the things they make is incredibly higher compared to other engineering firms, which have also let their engineering rot under the rule of MBAs but run with the advantage that their failures don't make headlines or get people hurt. More than understanding specific incidents or quality assurance, here I am more into understanding how and why designs stick for too long.

Consider one of the most iconic and widely used jetliners in the world, with a design history that spans over half a century. We can keep it rather anonymous, although you will quickly know what I am talking about. From its inception in the 1960s to the present day, this aircraft has seen numerous iterations, upgrades, and redesigns, reflecting the evolving needs of the aviation industry but also illustrating the reluctance of organizations to let go of a market winner.

The story begins in the 1960s, when this manufacturer, looking to expand its product line beyond its larger models, designed this aircraft as a smaller, more efficient short-haul airliner. The first model made its inaugural flight in the late sixties. It was soon followed by a stretched version with improved aerodynamics. These early models set the stage for what would become a highly versatile and long-lasting family of aircraft.

Decades later, the manufacturer introduced an updated series with advancements such as improved engines and greater passenger capacity. This generation solidified the aircraft's reputation as a reliable workhorse in both commercial and military aviation. In many cases, "workhorse" tends to be a euphemism for "a thing that pays the bills, so we better not touch it or breathe too loud next to it".

When the manufacturer decided to update the design with more fuel-efficient, higher-thrust engines, they faced a technical challenge. The new engines were larger in diameter than the originals. Thus, the aircraft started featuring flat-bottomed engines to keep the landing gear the same and maintain adequate ground clearance. This distinctive design, often referred to as the "hamster pouch", allowed the engines to be mounted closer to the wing, thus maintaining the required clearance without major changes to the aircraft's design. You have got to love it when a hack becomes a "distinctive" design feature. I have seen dirty hacks getting patented as novel inventions.

The next significant leap came with another series—called "Next Generation"—offering various improvements like upgraded avionics, more efficient engines, and increased range. This iteration was a commercial success, becoming a mainstay for airlines around the world. Another red flag: when something is called "Next Generation", you'd better start asking some questions.

The most recent chapter is the problematic latest series, introduced in the 2010s. This generation was designed to be even more fuel-efficient and environmentally friendly, featuring new engines and aerodynamic enhancements. However, it faced significant challenges. Two fatal crashes led to a worldwide grounding of the fleet. The tragedies showed issues with the aircraft's flight control software and prompted extensive investigations, modifications, and a rigorous recertification process.

Now, if we compare the latest iteration and the original model from the late sixties. Are they the same thing? It's always tricky to define the concept of "sameness" when we talk about technical objects.

"The Ship of Theseus" is a philosophical paradox that explores the nature of identity and change. It centers around a legendary ship owned by Theseus, the mythical king and founder-hero of Athens. As the story goes, the Ship of Theseus was preserved in a harbor as a historical artifact. Over time, its wooden parts began to rot and were replaced with identical materials. Eventually, every part of the ship was replaced, leading to a compelling question: Is the restored ship still the same Ship of Theseus? What is more, if the removed original parts were reassembled into a ship, which, if either, is the true Ship of Theseus? This paradox illustrates what it means for an object to remain essentially the same despite changes and renewals.

The manufacturer's strategy in incrementally redesigning this aircraft over the decades reflects a calculated intention to stretch an existing system to achieve a multi-factor goal: build on top of a winning product with an enviable market share, but also save the significant hassle compared to developing an entirely new aircraft. This approach was driven by economic and competitive factors, as well as the desire to meet the evolving needs of airlines without incurring the significant costs and risks associated with a new program, which could cost billions. By building upon the proven design of the original, the manufacturer was able to introduce new features, technologies, and improvements while maintaining a design "continuity". This allowed them to utilize many of the same parts, manufacturing processes, and even pilot training protocols from one generation to the next, reducing the time and cost associated with bringing these new models to market. This not only permitted a smoother certification process with aviation authorities, as the aircraft was already familiar, but also appealed to airlines, which could integrate these new models into their fleets with minimal additional training and without the need for extensive new support infrastructure. You can see how all this could make sense.

However, the inescapable nature of Diminishing Returns in engineering always kicks in. They have surely made solid revenues for decades, so I'm sure they don't regret a single thing, but it all indicates they seem to have gone beyond Theseus' limit, where the design is showing signs of obsolescence, and customers' trust has been impacted by a series of recent, high-profile negative events.

No solution lasts forever because problems evolve as a function of time. By accumulating small improvements, we only approach faster to the point where the thing is not the same thing anymore, but something else. Clever engineers start a new design before the previous one reaches unsolvable obsolescence, as we spoke in the previous section.

Engineers preach about reuse, but reuse is a double-edged sword. Bad things can happen when you mindlessly reuse a design in scenarios that have drifted from its original boundary conditions. All designs eventually become obsolete; the key is to recognize that the moment comes before reality does it for you.

Chapter 11
Fake Requirements

Abstract Ah, the tragic story of "the requirement", a concept that could have been a great tool for engineers to capture genuine stakeholder intent, but degenerated into a meaningless pose, where engineers write vacuous text for the solemnity of it long before they know what problem they're trying to solve.

Keywords Engineering design · Requirements · Project management · Verification · Work breakdown structure · Design process · Conceptual design

Sometimes, engineering can go on autopilot. And when autopilot kicks in, engineers start forcing things into existence for the sake of soulless compliance, pointless solemnity, or for appearances. One common victim is the requirement.

Requirements have become a robotic instrument used to create an aura of "serious work", only to end up being disconnected both from what the customer wanted and from what the design will end up implementing. Because, at the end of the day, reality will provide calibration down the road to adjust the latter with the former. Sooner or later, the design activity eventually adjusts to what the customer wants (because there are reviews, demos, and trials as the project evolves, and customers will surely express how they feel about it) while the original requirements will sit there in silence, stating things that someone came up with only for the sake of filling pages and complying with deliverables.

The way engineers sometimes handle requirements—faking them before they even remotely know what must be done—adds not just zero value to the engineering design process, but in fact, it brings a negative contribution, because you still need to input energy to write them and organize all those bogus statements, although they map to nothing whatsoever in the final design. It's all laughter until the time comes when you need to verify the system you have built against those invented requirements. Whoopsies. By then, the ones who wrote the fake requirements, if clever enough, have conveniently hidden somewhere safe and can't be made accountable.

Requirements done this way do a total disservice to the otherwise noble task of getting to know what your customer wants and coming up with a concept out of it. There is beauty in the first steps of a project aimed at designing a complex thing. The

I. Chechile, *Engineering is Broken*, https://doi.org/10.1007/978-3-032-18169-5_11

confusion, the ambiguity, the sense that everything is up for grabs. It is enjoyable to interrogate the stakeholders with questions and try to understand what they want. It is captivating to data mine their minds because it also teaches about communication (human communication) and how people express what they want, or what they do not want. There is a certain magic in the process of guiding customers and helping them converge into something, not without a dose of opportunistic nudging, so they end up choosing something that makes sense for you; that's the immaterial fee we can charge them for helping them figure things out, isn't it?

The joy is killed when bureaucracy steps in and asks for requirements when the data mining with the customer hasn't been done yet.

Here is a theory: the problem with requirements is not malice from engineers but just their incapacity to stop the managers from constantly disrespecting the natural flow of the design process. In that anxious sloppiness, the conceptual design stage and the requirements capture phase invert their positions in the lifecycle, and engineers start designing BY requirements instead of FROM requirements, using requirements as a makeshift sketching tool.

The right sequence is and should have always been: first, you talk to your stakeholders, you ask your questions, you do your homework, you understand the problem well, then you go and write down the curated list of things you need to do, so you and the team do not forget. And right after that, you start sketching. In the process, some items in your list might spawn "children" things that hang from them. That's what the solemn bureaucrats call "flow down" of requirements.

And if all this mess is not enough, some engineers lack the, how to call it, situational awareness to adjust the requirements creation activity to the problem at hand. If they happen to come from "classic" engineering environments, they will apply an incredibly heavyweight approach to capture requirements, whether they are designing a space shuttle or a picosatellite for a university. Both systems need to work, no doubt, but the intensity of the process for one or the other can't be just the same, can it? The way to create requirements can't be detached from the underlying complexity of what you're trying to do.

In short, requirements are not and should never be created for appearances or solemnity. Requirements must be analyzed into existence. Requirements are a glorified to-do list. In a healthy engineering environment, requirements are simply a summarized transcript of a conversation between you and your customer in a set of concise phrases. Nothing more, nothing less.

Do your bit. Don't be an accomplice to this stupid forgery. Kindly refuse to create requirements out of your hat. Understand what needs to be done. Don't use requirements as a design tool. And if a proposal (and funding) depends on your creativity to pull requirements out of thin air, make sure you don't take premature design decisions you will regret for years to come.

Chapter 12
Bad Work Breakdowns

Abstract Ideated decades ago in the context of big military projects, work packages today mostly emerge from guesswork. In many organizations, they are fabricated to meet proposal deadlines, etched in contracts, and treated as the ultimate truth. Work packages frequently lack connection to real Product Breakdown Structures (PBS), detaching planned work from what's being built. When work breakdown structures don't mirror the system's physical structure, problems appear everywhere, particularly in domains that cross-cut the system hierarchy. These issues point to a deeper lax definition of parts, wholes and taxonomies, leading to badly-defined hierarchies and a great deal of confusion.

Keywords Work package · Work breakdown structure · Product breakdown structure · WBS · PBS · Project planning · Deliverables · Program management

What an invention, the work package. An inescapable element of every project-based organization (especially government-related), the work package suffers from the same abuse requirements suffer from: if we don't have enough information to create them properly, we forge them into existence.

The modern work package was born in the heat of the Cold War. In the late 1950s and early 1960s, massive U.S. defense and aerospace programs faced a problem: how to coordinate thousands of contractors, millions of parts, and billions in budgets. The solution was the Work Breakdown Structure (WBS), formalized in 1962 through some DoD standards.[1] The WBS decomposed complex programs into smaller units, and at its lowest level lived the work package: the smallest chunk of work that could be planned, assigned, budgeted, scheduled, and verified.

Work packages were conceived as contract-anchored accountability units, each with scope, cost, schedule, deliverables, and acceptance criteria. This allowed project managers to track cost and schedule performance with a precision that matched the rigidity of those programs.

---

[1] For the bureaucrats reading: MIL-STD-881.

I. Chechile, *Engineering is Broken*, https://doi.org/10.1007/978-3-032-18169-5_12

As project management practices escaped the defense sector like radioactive monkeys from a laboratory and spread through handbooks, certifications, standards, and corporate project management offices, work packages were exported to other industries, adopting in the process the language, but often without the rigor that had made work packages originally useful. They turned from outputs of thorough, structured product decomposition into placeholders written long before the real system was even remotely understood.

Hierarchical work breakdowns are quite inevitable in our craft due to the fact that we tend to create hierarchical systems, and thus the work needs to be organized in a way that's more or less aligned with the architecture of the thing under design. Despite its incredibly boring name and appearance, the work package by itself is not a terrible invention; the problem lies in forging them. The mechanism goes like this:

> Boss: Put together some work packages in the proposal, come on, the deadline's on Monday!
> Subordinate: But boss… we still don't know what needs to be done.
> Boss: That doesn't matter!

And off they go, pulling rabbits out of hats with their little boxes in the diagram. As I write this, someone, somewhere, is sitting down eliciting work packages out of thin air: WP1100, WP2100, WP3100, followed by the inevitable tidal wave of zombie deliverables in the form of eternal documentation.

Those innocent work packages, written when nobody really knows what the heck needs to be done, end up carved in stone for the entire duration of the project because changing them later is usually seen as a bad sign. In certain engineering circles, changing what your proposal said triggers a wave of contract amendments, penalties, meetings, nervous calls, and a letter to the King.

Work packages can make sense when done right: breaking work down into discrete chunks that are contained and manageable. Who could be against that?

But a work package needs to be anchored to something real. It can't just come flying out of a hat. A work package must originate from a Work Breakdown Structure (WBS), which in turn must originate from a Product Breakdown Structure (PBS), which is what reflects the physical architecture of what we're trying to build, which in turn must reflect what the customer wants. If the needs are not understood, the PBS is ill-defined, the WBS will be ill-defined, and therefore, all work will be ill-defined, making everybody miserable in the process.

And since artificial superintelligence hasn't yet reached the point of designing things on its own (we talk about that later on), engineering is still largely a human affair. Which means that people and how they are grouped are an essential part of the work breakdown activity.

When engineering teams are aligned with the physical architecture of the thing being designed and built, we're all in Disneyland since this creates little to no ambiguity, and it's easy to map work packages to teams and to physical system elements. But when they're not—for example, when one subsystem is spread across multiple physical components owned by different teams—welcome aboard to the hell that

Mel Conway[2] predicted more than half a century ago, and we still don't listen to the wise man.

This is where mediocre project managers and systems engineers mess it up. And no certification will save you here, because it's a matter of common sense and experience, not credentials. If you create work packages that ignore boundaries—technical and human—and don't map to anything concrete, don't expect anything but confusion. This problem gets particularly itchy when creating work packages related to abstract categories or overarching domains that touch practically everything. The classic example? Software. In a complex system with multiple subsystems, there's probably software living in more than one place. And that's when everything goes up in flames, because those work packages are now mapped to multiple groups and subsystems and end up like children of divorced parents: a bit here, a bit there, conflict, arguments, and years of therapy for everyone involved.

Once again, I find myself asking us to go back to the sources and use things as they were meant to be used. Let's stop treating work packages as magic square boxes that legitimize a project or a contract simply because they exist, just like we do with requirements.

We have to stop faking work packages before the system exists, and we must let them fall out of the architecture. Rather than writing arbitrary WPs in the proposal stage out of someone's creativity on a random day, the scope should emerge from a well-defined Product Breakdown Structure once it reaches minimum stability. Work is then structured around real subsystems, components, or interfaces, and not around guesses. All while keeping an eye on team structure, as Mel Conway would have wanted it. This way, scope evolves in iterations tied to architectural maturity. Outline the system first, then define the work.

12.1 Confusing Parts and Wholes

Perhaps the most basic mental trap when it comes to breaking down engineering work is not totally about faking it for compliance (which we do), but about not knowing what goes inside what, what is a whole, or what is a part; what's a parent, what's a child.

Now, some complicated terms. A meronomy is a type of hierarchy that deals with part–whole relationships, in contrast to a taxonomy whose categorization is based on discrete sets. The study of meronomy is then known as mereology. For example, in linguistics, a meronym is the name given to a constituent part of something. For example, "X" is a meronym of "Y" if an X is a part of a Y.

[2] Conway's law describes the link between communication structure of organizations and the systems they design. It is named after the computer scientist and programmer Melvin Conway, who introduced the idea in 1967.

Many of the kinds of artifacts we create as engineers are things composed of other things. So, engineers work a lot with meronomies, although they probably haven't heard the term before.

Poor, half-baked part-whole definitions cause a lot of trouble. This is an incredibly underrated problem in engineering that transpires as perhaps one of the most pervasive pitfalls in everything we do. We structure things badly, simply because we think about them badly.

As we discussed in the previous section, a great example of the part-whole confusion is software. In complex systems, software is always a *part* of a containing whole. For example, an acquisition board that is inside a unit in an avionics bay in an airliner; software resides inside the board. Software is a *meronym* of the board. In cyber-physical systems of a reasonable complexity, software is a component, as it lives inside something that resides inside something.

This confusion causes a lot of trouble, and trouble again maps into the organizational level, basically because both the system and the organization that designs the systems are just one, deeply entangled thing. It's always interesting seeing someone being christened as "Head of Software" or similar in an organization that sells systems. Good luck resolving the problem of leading a portion of everything that contains software and nothing really at the same time. It's like being promoted to "Head of Screws and Washers". What jurisdiction will you oversee when there are screws everywhere in the system? Is the whole system your jurisdiction, then? Of course not.

Truth is, we suck at meronomy, and the worst part is that we don't even know we do. We are fully convinced that a feline is a type of tiger.

12.2 The Tree of Truth (Product Ontologies)

After discussing breakdowns, meronyms, parts, wholes and what have you, we need to conclude this topic with the inescapable and to some degree mysterious concept of ontology. The term "ontology" carries an air of philosophical sophistication that tends to make engineers uncomfortable. It sounds like the sort of thing that requires a fondness for metaphysics and the nature of existence. However, if we strip away its academic aura, ontologies address a very practical problem every engineer knows: keeping track of what things are, how they connect, and where to find information about them.

I get so worked up by the Digital Transformation discussion that, if someone really, truly wants to talk about a true digital transformation, we need to talk about ontologies and how they're the backbone of any dreams of tying the physical with the digital, and how we barely even talk about them. At the end of this section, I hope you can get to see how far we are from such a transformation. Mind you, this might get a bit detailed.

Software engineers have been doing ontology engineering for decades without calling it "ontology engineering" but instead calling it "object-oriented programming". Classes define types of things, inheritance establishes taxonomic relationships, properties capture attributes, and methods represent behaviors. The main difference is that OOP ontologies live inside code and serve computational purposes, while formal ontologies exist as explicit knowledge structures across system boundaries. If you can design a class hierarchy, you can design an ontology.

Granted, *Ontology* sounds a bit presumptuous, doesn't it? *Oh là, la* ontologies. But we are not Greek philosophers, we are engineers, so let's simplify this.

Ontologies are not philosophy, but pure computer science, and an essential tool to represent knowledge in a way computers can process it. An ontology is ultimately a "tree of truth" of sorts,[3] whose trunk, branches and leaves reflect relationships between things relevant to a given context or entity. We all carry our own "trees of truth" wherever we go; we have attributes (height, weight), we are related to people and objects, we are composed of constituent parts, we perform functions, so we actually could capture all that information in a tree-like structure that would describe us completely. And the same applies to any other object,[4] including technical artifacts. And the fact is, such a tree of truth always exists, whether we make it explicit or not.

When a maintenance technician asks which sensor monitors compressor-exhaust temperature on a gas turbine, they are performing an ontological query in the gas turbine's tree of truth. In most engineering organizations, this tree is never formalized, so answering straightforward questions requires consulting disconnected systems and relying on people's memory, including that one grumpy veteran who somehow always knows.

Ontology literature emphasizes domain ontologies—general concepts and relationships—but an earthlier and more practical use lies in product ontologies. Domain ontologies capture general knowledge ("all pumps have motors"), while product ontologies model specific physical artifacts at the instance level once the object exists. This makes them concrete representation models of assets. Instance-level ontologies are verifiable: the object exists, can be observed, measured, and documented. Boundaries are explicit, properties are concrete, and relationships can be validated directly. When modeling a specific gas turbine, there is no ambiguity about whether a blade is part of the assembly; you can check the physical object or its schematics.

Physical artifacts exhibit natural hierarchical decomposition aligned with ontology structures. An aircraft decomposes into fuselage, wings, and propulsion; the fuselage subdivides into frames, stringers, and fasteners. This part-whole hierarchy forms the backbone of product ontologies, but relationships extend beyond containment. Objects manifest structural (*connectedTo*, *mountedOn*), functional (controls, powers, cools), spatial (*adjacentTo*, inside), causal (causes, *affectedBy*), and temporal

[3] Another way of calling ontologies is Knowledge Graph Schemas (I am aware the plural of schema is *schemata*, but the incorrect way is so popular that I prefer to use it).

[4] It also applies to this book, whose tree of truth connects all concepts, ideas, and chapters by relevant relationships.

(precedes, *replacedBy*) relationships. This multiplicity allows a product ontology to unify perspectives on the same artifact. An aerospace pump is simultaneously a mechanical assembly, a process equipment item, a maintenance asset, and a software-controlled device. The ontology provides a coherent representation across these views.

An essential characteristic distinguishing product ontologies from just a conventional database is their role in linking physical objects to multiple digital representations of them. Any engineered artifact exists not only as a physical entity but also as CAD models, schematic symbols, simulation models, sensor time series data streams, maintenance records, and documentation. The product ontology must track the identity of the physical object and relate all possible representations to it in the tree of truth. Consider, again, a gas turbine compressor blade: the physical blade has a serial number, material composition, installation date, and location. Its CAD representation exists as a STEP file with geometric and manufacturing information. It also appears as a simplified symbol in process flow diagrams. Finite element analyses model its stress distribution under operational loads. Vibration sensors monitor its condition in real-time. Inspection X-ray documents its integrity condition. Maintenance records track repairs and replacements. Without a structure organizing and connecting it all, these informational nuggets exist as isolated information fragments scattered across systems. The product tree of truth could bind these disparate representations to the singular physical artifact, allowing us to query through representation boundaries. This kind of data integration proves particularly valuable for retrieval systems operating across different engineering informational domains. A query about component failures should simultaneously retrieve relevant CAD models, stress analyses, in-field sensor anomalies, maintenance histories, and design documentation. In the way we currently do this, all these reside in separate information silos, which requires humans to manually sift and tie the knots across multiple types of data. Product trees of truth, well done, could enable a semantic linking that would pay off so much: a user asks about turbine blade failures, the system identifies the relevant blade instances in the ontology, retrieves all linked representations, and presents a coherent, multi-modal response. The tree serves as the semantic glue binding otherwise disjointed information. Today, we human engineers are the semantic glue, and as information grows in complexity and size (along with the complexity of the things we create), there will be a point where this analog approach won't scale.

Product ontologies can be constructed and fed from existing engineering data sources. The infamous Product Lifecycle Management and Product Data Management systems—that we kind of scold thoroughly in Sect. 15.2—are sources of invaluable data, because they typically maintain bills of materials, assembly relationships, part revisions, and associated documentation. APIs to these systems—when they don't crash on us—would allow programmatic extraction of part hierarchies and configuration data, all important ingredients of any respectable tree of truth. Each part number then becomes an ontology class; hierarchical extraction in the BOM becomes partOf relations in the tree. Revision histories can provide temporal relationships. Designators extracted from drawings become instance identifiers.

An inescapable challenge is that physical objects evolve through their lifecycle, and a product's tree of truth must shadow this temporal evolution. As designs mature through revisions, CAD models move across versions. The ontology must maintain relationships between physical objects and the correct revision of their digital representations, ensuring that retrieved information corresponds to the actual as-built configuration rather than outdated design revisions.

The implementation of a product's tree of truth can face some challenges. Scale represents a true concern; complex systems contain thousands or millions of components. A commercial aircraft comprises millions of parts. Ontology size impacts query performance and maintenance burden. As in any other modeling activity, granularity decisions are critical. Should every bolt appear as a distinct ontology instance, or should fasteners be aggregated? The answer depends on maintenance requirements, traceability needs, and analysis objectives.

Data quality and consistency present challenges as well. Ontologies informed by as-built documentation reflect actual installations, including deviations from design intent, field modifications, and documentation gaps. Reconciling multiple information sources requires conflict resolution strategies and validation workflows. Automated extraction always introduces errors requiring human review.

Machine learning purists might find trees of truth somewhat itchy, because methods in that domain, especially in deep learning, embrace a "learn from data" philosophy where the model discovers patterns without imposing rigid structure, while ontologies represent a more "rationalist" approach of encoding explicit knowledge hierarchies. These approaches often feel at odds. Modern success of machine learning comes (partly) from learning representations automatically rather than handcrafting features, so ontologies can feel like a step backward toward the manual knowledge engineering that machine learning was partially designed to escape.

Building good trees of truth can be time-consuming, requires domain expertise, and needs ongoing maintenance, and in fast-moving projects, this isn't always justified when the usual approach works reasonably well without it. The problem with this is that this usual approach requires humans acting as the semantic glue across disjointed systems and pieces of information, and as the complexity of the information created grows rapidly (tied to the increasing complexity of the systems we engineer), the approach will eventually (read, soon) stop scaling.

Chapter 13
Fake Ownership

Abstract Technical ownership is supposed to define those who can meaningfully steer the evolution of a manageable unit of engineering work. Modern frameworks preach on the role of the Product Owner as a single accountable figure managing goals, backlogs, and stakeholder priorities. In practice, artificial, imposed ownership clashes with natural ownership that arises from those who design and deeply understand the object of interest. This means artificial owners lack the technical insight to navigate the dependencies, making their authority superficial and their influence limited to administrative backlog maintenance.

Keywords Technical ownership · Product owner · Backlog management · Artificial ownership · Natural ownership · Engineering management · Product development

The concept of ownership in engineering is as itchy as it gets. I am not talking about ownership in the literal sense of the word (possessing something) but about technical ownership, epitomized by the concept of the "Product Owner" as the Agile deities have come to call it. The Scrum Guide defines a Product Owner this way:

> The Product Owner is accountable for maximizing the value of the product resulting from the work of the Scrum Team. How this is done may vary widely across organizations, Scrum Teams, and individuals. The Product Owner is also accountable for effective Product Backlog management, which includes:
>
> - Developing and explicitly communicating the Product Goal.
> - Creating and clearly communicating Product Backlog items.
> - Ordering Product Backlog items.
> - Ensuring that the Product Backlog is transparent, visible and understood.
>
> The Product Owner may do the above work or may delegate the responsibility to others. Regardless, the Product Owner remains accountable.
>
> For Product Owners to succeed, the entire organization must respect their decisions. These decisions are visible in the content and ordering of the Product Backlog, and through the inspectable Increment at the Sprint Review.

I. Chechile, *Engineering is Broken*, https://doi.org/10.1007/978-3-032-18169-5_13

> The Product Owner is one person, not a committee. The Product Owner may represent the needs of many stakeholders in the Product Backlog. Those wanting to change the Product Backlog can do so by trying to convince the Product Owner.

Some things from that excerpt above make sense. Every product under development should have clear goals and a planned evolution. A bucket list to keep track of all that doesn't hurt, does it? You want to call it a bit fancier than "bucket list" and call it "backlog"? Ok, backlog it is.

Equivalently, every product needs a "funnel" between business and engineering, filtering in the process what makes sense to add—or to remove—versus what does not. All in all, the backlog seems like a good thing, and if the so-called Product Owner can both handle the backlog and the funnel, that's actually not bad.

But here's a fundamental problem: there are two kinds of ownership when it comes to technical artifacts. There is an artificial ownership and a natural ownership. With artificial ownership, I mean a fictitious, imposed kind of ownership, where a higher-up unilaterally assigns someone to be the *owner* of thing X. This arbitrary ownership collides with the other kind of ownership: the natural ownership. Natural ownership is exercised by those who have developed and shaped the object from its conception and therefore know it by heart. We all feel ownership of our own creations.[1]

And there lies the problem; the problem of imposing artificial owners with shallow knowledge on the matter and having them try to exercise their hollow authority, only to find that they own nothing in reality and instead are consistently sidelined by the natural owners who have the domain knowledge to call the relevant shots for modifying, adding or removing features.

Artificial owners cannot really drive the Backlog besides doing bookkeeping stuff or shuffling some inoffensive things here and there, because any serious rework would require navigating through the dependencies and intricacies of the tasks involved, which are naturally owned by the domain experts. Once in a meeting, a Product Owner asked if a board layout could be pushed earlier than the schematic capture for the board, to the general giggling of all electrical engineers around the table.[2] And that's that; when the Product Owner asks such a silly thing, revealing they have no insight into the underlying process they are supposed to "own", the respect is lost forever, and there is usually no way back.

In most cases, artificial owners just become part of what a former boss of mine (J.R.) would call the "useless layer", and are left as simple representatives of the interests of certain stakeholders, tasked with the ungrateful endeavor of negotiating with the natural owners when and how certain features can be moved up or down in the pipeline. Artificial owners are destined to be simple backlog maintainers whose main persuasion technique might be the best puppy eyes they can display to get something out of the real owners.

Maybe it's a naming thing. Had the Agile fathers named these roles in a bit less "grabbing" way, things could have been maybe a tad different. But no, everyone must be a master, an owner these days. Had the Product Owner been called Product Facilitator or Product Helper, their existence could have been a bit easier.

---

[1] We also feel ownership of the creations of others; this is called the IKEA effect.

[2] In a professional capacity, board layout always comes after schematic capture.

Chapter 14
Minimal Engineering

Abstract Is it possible to do *too much* engineering? Is there a point beyond which doing more of it doesn't pay off? Certain problems in engineering scale directly with the resource you throw at them, which means adding more of that resource doesn't solve the problem; it only makes it worse. In fact, too much of *anything* is bad. Yield at the outputs will inevitably start decreasing sooner rather than later as you keep on pumping the inputs, all other things equal. The law of diminishing returns is inescapable and is at the heart of the engineering craft.

Keywords Diminishing returns · Yield · Engineering process · Organizational growth · Minimalism · Moderation · Goldilocks engineering

Certain problems in engineering scale directly with the resource you throw at them, which means adding more of that resource doesn't solve the problem; it only makes it proportionally worse.

For example, take the problem of power consumption in microchips: if power consumption is directly proportional to the number of transistors integrated on the chip, you can never expect to solve it by adding *more* transistors, can you? Now see the issue of space debris: the more satellites there are in orbit, the likelier to cause a Kessler Syndrome, where a collision would increase the probability of more collisions, creating a chain effect. You can never expect to solve the space debris problem by launching more satellites.

Organizational growth shows a similar effect. With more people, there is a combinatorial increase in the number of people's interactions; therefore, the coordination and communication overhead increase proportionally. More people, more noise. When the overhead and noise become a problem, you can never expect to solve it by bringing in more people. This is nothing else than Brooks' law at play. Brook's law is an observation about software project management according to which "adding manpower to a late software project makes it later". It was coined by Fred Brooks in his 1975 book *The Mythical Man-Month.* Reads obvious, but still, it's a mistake made over and over.

I. Chechile, *Engineering is Broken*, https://doi.org/10.1007/978-3-032-18169-5_14

Now, an immensely missed point for many engineering circles is that the inverse of the effect is not necessarily true. If the amount of a resource is the main problem, dramatically reducing it will not magically solve all the issues. It can, in fact, create new, uglier problems. Cutting the people in a project in half may reduce coordination overhead and cost, but who will do the work that the missing people were doing? (assuming they were doing something). Who will work on redistributing responsibilities in a balanced way when all of a sudden half of the souls are gone? The solution to a Kessler Syndrome is not deorbiting satellites to make orbits less crowded; this would disrupt essential services satellites provide in our everyday lives.

Growth brings proportional pains, but a mindless guillotine of such growth brings new pains. We as engineers must carefully navigate this tradeoff, measuring twice but cutting once.

Then, is it possible to do *too much* engineering? Is there a point beyond which doing more of it doesn't pay off? The answer to this is quite simple: why wouldn't an activity where complexity and the use of resources are at the core of it show the same dynamics as in other domains where processes yield progressively smaller improvements as the easiest gains have already been captured, and what remains is harder, slower, or more marginal to improve? So, that's a long paragraph to say: yes, there can be too much engineering.

Seen as a black box, engineering can be thought of as a transform function. It takes some input, and it provides an output, which is the result of its internal activity. A mechanistic take, sure, but not necessarily untrue. Granted, it is not a single-input–single-output black box, but instead has multiple inputs and outputs, with cross couplings galore, and shows complex adaptive behavior. But, if you stop feeding the box with what it consumes, sooner or later, it will stop producing anything at the output. This is obviously applicable to many things in life: A company without funds, a restaurant without customers, a plant without water.

A flawed logic that seems to reign around many minds of the engineering kind is that, for every increase at the input, you should always obtain an equivalent increase at the output. As if the engineering activity will not show any opposing dynamics to these ever-increasing stimuli. Well, that is clearly not the case.

In engineering, diminishing returns manifest when additional investments of time, effort, or resources yield progressively smaller improvements in performance or reliability. Or even more, when adding those factors makes everything worse. This phenomenon also shows in design optimization, where initial design iterations deliver substantial gains, but subsequent refinements produce marginal benefits relative to cost and complexity. In fact, too much of *anything* is bad. Yield will inevitably start decreasing sooner rather than later as you keep on pumping the inputs, all other things equal.

There will be a point beyond which bringing more people will not improve the schedule. There will be a point beyond which more Systems Engineering will not benefit your project cost and actually increase it. There will be a point beyond which more testing will not improve software quality. There will be a point beyond which more features will not make the product any better. There will be a point beyond which more tools will not improve a process any further. There will be a point beyond which more documentation will not make things any clearer.

14.1 Minimalism in Engineering

As engineering practitioners, we see time and again that the relationship between simplicity and complexity often defies logic. Engineers design systems according to rather straightforward principles, only to watch them behave in ways no analysis predicted. Meanwhile, problems that appear hopelessly intricate sometimes converge to solutions of shocking simplicity.

A microchip implements small logic gates, each performing trivially simple operations, yet billions of these gates working in concert create systems capable of tasks that seem almost magical. No single transistor *understands* anything, yet their collective behavior enables things like deep learning, real-time language translation, and simulations of entire universes. This complexity emerges not from sophisticated components but from the sheer number of simple interactions happening simultaneously.[1]

A programmer writes (or tells an LLM to generate) modular code where each function performs one well-defined task. The individual pieces make perfect sense. However, when these functions call each other, when they share state and pass data back and forth, when they interact with external systems with noise, delay, and user inputs, bugs emerge that can take a long time to trace. A system composed entirely of correct parts can still behave incorrectly because complexity arises in the interstices between components, in the interfaces and interactions that no single piece of code controls.

Complex failures sometimes trace back to absurdly simple causes. We all know the case of the Mars Climate Orbiter and how it went terribly wrong because one team used metric units while another used imperial. The spacecraft itself represented extraordinary engineering complexity, but it failed due to a simple unit conversion.

This understanding should reshape how we approach design. Our reflex as engineers is that, when facing a complex requirement, we create an equally complex solution to match sophistication with sophistication. Frequently, the most robust systems emerge from simple, well-understood components arranged cleverly. Premature complexity is a big problem in our practice.

The challenge is in sizing the problem we face correctly. Some engineering problems genuinely require intricate solutions because they operate in domains with irreducible complexity. Designing a gas turbine or a nuclear reactor demands sophisticated engineering because the physics involved doesn't simplify easily. However, many problems that appear complex are actually emergent phenomena arising from

---

[1] This is the central idea in *connectionism*, which states that a large number of simple computational units can achieve intelligent behavior when networked together. This insight applies equally to neurons in biological nervous systems and to hidden units in computational models.

simple systems, and adding more complexity to the solution only increases the unpredictability.

The most elegant engineering involves finding the minimal solution that leads a system toward desired behavior, rather than trying to control every single variable. Anyone who has ever engineered anything has felt that magical dopamine shot when an incredibly simple solution solved a complex problem that seemed unsolvable.

14.2 Intuitiveness and Idiot-Proofness

And now that we are discussing simplicity, let's also squeeze in a short discussion about intuitiveness and fool-proofness as values that engineering seems to have lost, or at least de-prioritized.

I'm an idiot. I assume those who are (still) reading these lines had a firm suspicion, but today I'm confirming it. I am. And as the self-aware idiot I am, I demand to be treated as such. Anything that requires my attention must be strictly designed, written, and explained in an idiot-proof manner.

The older I get, the less I seem to understand. If I live long enough, I expect there will come a day when I won't understand anything at all. Today I understand less than yesterday, and a bit more than tomorrow. Lately, I read any instructions, and in most cases, they don't make sense. However, in my defense, everything instructional I read lately doesn't lead me toward an "immersion in the problem". I need to *dive* into something to understand it; from the surface, nothing really sticks. It needs to click for me. And it's rare to find things that manage to place me in that mental state where you suddenly connect with whatever the hell you are looking at.

Same story with everything else, besides reading manuals. I use a product, and I don't get it. Packaging? By the time I finally understand what that industrial designer wanted to impress some boss with, the thing is already torn apart beyond recognition. Ohh, *that* tab was supposed to fold *that* way. Meh.

My son had a phase when he was obsessed with Transformers, and good Lord, the time it used to take me to figure out how to turn the little robot into a car or back again. A click here, a twist there, rotate something that shouldn't rotate—and all with the kid staring at me, eyes full of hope—until bam, the robot breaks, and you're left wondering how to explain that this toy is never going to be the same again.

Writing instructions, creating products or interfaces in idiot-proof form, is a rare virtue. The creator must let go of the urge to look smart or superior and accept that most of the people interacting with their creation will very likely be idiots like me. And that ease of use is the highest compliment a designer can ever receive.

If achieving that means over-explaining, then over-explain. It's better for a user to dismiss redundant instructions than to feel that existential void of not having enough information to understand how to interact with a technical artifact. But careful: designing idiot-proof stuff doesn't mean insulting or underestimating the user. It only means respecting them enough to never leave them alone in a sea of ambiguity. Designing for fools is designing for the hurried, the person who didn't read, the person who interprets everything backwards.

At the end of the day, you don't design for a hyper-alert, deeply engaged user who cares about your thing as much as you do. You design for the tired, the distracted, the person who just wants to get through this because their mind is somewhere else entirely. Many engineering designers need to get off their high horse and accept that they are not the center of the universe. Design for humans as they are, with flaws and all.

14.3 Nordic Moderation and Goldilocks Engineering

To finish this chapter about minimalism and diminishing returns, let's take a look at some Nordic wisdom. *Lagom* is a Swedish and Norwegian term that means: "in its right measure", or "just the right amount". Lagom is about finding balance through moderation: take just enough of what you need to do well without going overboard. The Swedish proverb, "Lagom är bäst", literally means, "The right amount is best" but is also translated as "Enough is as good as a feast". Engineers must always remember that there is virtue in moderation and think of lagom when designing, writing requirements, documentation, or code. The right amount is best.

Also, engineers can borrow concepts from children's tales. In the classic tale "Goldilocks and the Three Bears", a girl wanders into a bear family's home, tests their porridge, chairs, and beds, finds one of each "just right", and is discovered by the returning bears, prompting her to flee. The story illustrates the importance of proportion and the idea of an ideal middle ground. The Goldilocks zone is where things are *just* right; the sweet spot between too hot and too cold, too complex and too simple, too cheap and too expensive.

Chapter 15
Shitty Tools

Abstract Why do bad tools spread so easily? Whether an in-house abomination that shouldn't have seen the light of day or a commercial, expensive bag of bugs with abominable UIs and cryptic error messages, bad tools seem to spread quicker than what logic would deem reasonable. There is something especial about badly conceived tools. The worse the tool, the faster it spreads, and the more strategically it positions itself as the cornerstone of an architecture that would collapse without it. Before you know it, the bad tool becomes irreplaceable. Can it get worse? Of course, and that's when bad tools reach enterprise level.

Keywords Bad tools · Tool adoption · User interface · Enterprise software · Software migration · PLM · ERP · CRM · Software migration

All it takes is a walk through an airport and a glance at any open laptop to see that, out of ten people you peek at, ten are writing an email, working on a document, using Excel, or making a presentation in PowerPoint.

An airport is, without a doubt, a demographically heterogeneous environment when it comes to professions: there must be doctors, lawyers, engineers, and marketing people in roughly similar proportions; the place doesn't favor any particular group.[1]

One can safely say, then, that the use of these tools cuts across industries and occupations across the entire spectrum.

Like it or not, the reality today is that practically no job escapes the gravitational pull of software tools; tools we use every single day for replying to anodyne emails, writing bland documents, doctoring budgets in spreadsheets, and presenting lousy ideas in slides.

The undeniable fact is that as humans, we create tools, and in turn, tools shape us. When we use tools, the tools affect the way we think and act. Today, it's very likely that when we want to communicate something to an audience, we already think in "slide" format before even opening PowerPoint. At least I do. Similarly,

[1] With the exception, perhaps, of management consultants.

I. Chechile, *Engineering is Broken*, https://doi.org/10.1007/978-3-032-18169-5_15

when we want to reach out to someone outside our organization, we naturally think in "email" format before typing a single word, mentally outlining the intro, body, and the obligatory closing pleasantries.

One interesting shift in this era is that the tools we use are no longer so generic in the way physical tools once were. Back then, when someone needed a hammer, practically any hammer could get the job done. Tools within the same category used to be more or less the same across vendors.

Today, tools are far more specific. When someone needs Excel, they don't just mean any spreadsheet software; they mean that specific tool made by that specific vendor. Handing them LibreOffice isn't a viable solution because the learning curve would be too steep and time-consuming to ride. User interfaces and intrinsic tool complexity add to creating all these barriers; once you learn the tricks and shortcuts from a tool, you don't easily feel like learning it all again from another vendor, as good as they might be.

Email still is a bit like the hammer: all we need is to send and receive messages, write, draft, create folders, and filters. Does it really matter if it's Outlook, Gmail, or whatever else? Not much. The UI is similar enough, and the complexity low enough, that switching from one to another wouldn't be the end of the world.

However, more complex, configurable tools make us hostage. The time invested in learning them creates high mental barriers that make us nauseous at the sole thought of switching. Tell an electrical engineer they have to ditch their favorite EDA tool for an alternative one and watch the terror in their eyes. Or ask a programmer to give up their favorite IDE.

The more complex and convoluted the tool, the less we'll want to let it go, no matter how brilliant the alternative may be.

It goes without saying that knowing by heart certain tools also makes us belong. If you want to play the outsider and refuse to use Word, Excel, or the most popular IDE, good luck finding a job. Knowing these tools makes us part of the conversation because they're used by large audiences and mastering them becomes a ticket to fit in.

It's shocking how frequently the tool–user dynamics is underestimated by organizations that decide to build in-house contraptions that aim to replace incumbents for the most half-baked reasons ever. I have seen entire software teams chasing the marvelous idea of creating an in-house replacement for… Google Calendar. The result? Years of development to come up with something notoriously inferior to what it came to replace.

It's quite a *Chad* move when companies try to make a business by deciding to pursue, as their main product, a replacement for a massively popular tool. It's really hard to get people to switch, as if it's like swapping one hammer for another. I mean, bless them; old and bad tools need someone challenging them, and things can always be improved, but the quest is massive, considering the switching barriers are sky high.

Terrible tools work flawlessly in only one single aspect: as excuses. Mountains of incompetence and laziness can be comfortably hidden behind a bad tool. The worse the tool, the more you can blame it for your own inefficiencies. "I tried, but the tool crashed", or the tool won't start, or the tool corrupted the data. It's the modern version of the dog that ate the homework. I must admit I use the excuse extensively. Every time I have to hang a frame or put up curtains at home, and as I stare at the mediocre result of my work, I tell my wife that it's not me who is useless as a handyman. It's because I have bad tools.

There is, though, something contagious about badly conceived, badly designed tools. With their primitive user interface, painful configuration process, and an abundance of obscure text files or XMLs of mysterious and varied schemas. With zero user-experience consideration, packed with cryptic error messages like "error – 578". Crashing in our faces without the slightest warning. In-house bad tools beg for a special mention.

The shitty in-house tool is born by accident. It starts as an innocent experiment by someone. But then they show it to someone else, and that person says, "Mmm, I could use that". And then they show it to someone else who also says, "Huh, that could come in handy for X or Y". And before you know it, the abomination spreads like a fire. One day, you wake up, and it's everywhere.

Suddenly, nothing can be done without the shitty in-house tool, because it has become the central piece of the puzzle. All roads lead to it.

The shittier the in-house tool, the faster it spreads, and the more strategically it positions itself as the cornerstone of an architecture that would collapse like a house of cards without it. Before you know it, the shitty internal tool becomes irreplaceable.

The shitty in-house tool is unmaintainable because whoever built it years ago is usually long gone. Naturally, there's no documentation except for a shady README or some ancient.doc file made in Word 97.

Of course, shittiness is not an exclusive trait of in-house experiments, though. There are also terrible tools we pay for, and we pay a lot. All made by multi-billion-dollar companies that have massive development teams, only to sell tools that are fond of throwing an abnormal termination in your face in the worst time possible and have user interfaces designed by the enemy. But hey, please send feedback if you want that feature to be added in a future release!

The commercial engineering tool market has a tendency to develop monopolies or duopolies, where dominant players offer products that are despised by their captive users. These companies tend to host online forums that are populated with people sharing workarounds. Their documentation is useless or nonexistent. Things that should take seconds require countless nested menus and confirmation dialogs. Their software has a tendency to crash at precisely the moment you remember you were supposed to save. But everyone uses them. Why? Because, as we said before, the switching costs are astronomical. Because files are in weird proprietary formats. Because everyone else in your industry uses it. It's all a perverse collection of economic and technical moats that fuels their revenues.[2]

And the monopolists know this, oh yes, they do. They've learned their software has become irreplaceable and turned that knowledge into their main business strategy. There's a big difference between great software and irreplaceable software. The former delights users, makes hard things easy, and feels like it pays for itself. The

[2] Do you work with FPGAs? Geez, I can only feel sorry for you.

latter only needs to be good enough that the pain of staying is marginally less than the pain of leaving. They economically exploit the inescapable *better the devil you know than the devil you don't*. You feel dirty when you pay the yearly license fee.

Users of monopolistic software tools tend to cope. Learn workarounds. Share unofficial scripts that fix what should have been fixed by the vendor's dev team five years ago. Write the tutorials that the company should have written. Users of monopolistic tools become, in effect, unpaid R&D for companies already making millions from their suffering. It's quite a perverse cycle. And when someone finally builds something better, the monopolist either acquires them or leverages their market position to crush them.

Somehow, we have normalized this tool shitshow. We have accepted that professional tools can be hostile to us. That they can freeze, crash, corrupt files, ruin our work of days and months, and still dominate their markets. Something feels extremely odd about this dynamic, and we should stop pretending it's not.

15.1 The Enterprise Software Shitshow

There is a downside to complex product development, and it usually does not become visible until it turns into a ticking bomb: designing and producing technical objects generates a lot of data. Such data is usually called Product Lifecycle Data; in short, all the data created (and needed) to consistently plan the development and production of a technical product.

Product Lifecycle Data, despite the glamorous name, tends to be an inelegant, heterogeneous collection of spreadsheets, EDA/CAD files, code, scripts, documents, and what have you. When things are small, each department usually does its own housekeeping of the data for the designs it owns, and demand for this data outside the department tends to be rare. For example, bills of materials and inventories might be kept and maintained by a design engineer who also talks directly to the suppliers. But when things start to get bigger (in terms of system and organizational complexity), a supply or procurement department may pop up, and all of a sudden, those bills of materials and inventories become a valuable as gold, as well as the relationships with suppliers.

As growth accelerates, the budgeting activity (once an exercise of good faith between engineers) calls for a more professional and dependable approach, and the financial folks become very curious about how cost breaks down and build up. To make things even more colorful, product variants appear, different versions of parts or components appear, and all this means that now someone or something has to keep track of what goes where, what version, what variant, and what change has been introduced.

Suddenly, there is this surge of demand for product design and production data, and a pressing demand for cross-department coordination and configuration control across variants. All this puts a lot of pressure on the spreadsheets that are still heavily used at this stage, containing an increasing number of obscure macros and convoluted formulas.

Then, in an otherwise routine meeting, someone drops the bomb:

– You know, I think we need an ERP[3]

Or even scarier:

– We need a PLM[4]...

This moment tends to be neglected in the history of organizations, but it marks the end of the company's age of innocence. It is the critical moment when it's finally understood that the organization cannot go on by hoarding loose chunks of information all over the place, and enterprise software makes a triumphant entry.

The enterprise software question is perhaps the grandmother of all the mothers of rabbit holes. Doing it the wrong way can seriously damage a company. It contains all the ingredients a proper rabbit hole must have: vendor lock-in, learning curves, cross-department "pissing contests", and thousands of hours of error-prone data migration done awfully manually by external parties who don't understand the context and will charge you a good price for doing such a torturous task with subpar quality.

When selecting enterprise software, too many things must fall into place. First, the ability of the enterprise software to grow with the business, or how the needs will evolve when the company doubles or triples its size. Also, the ease of integration between different building blocks of enterprise software: ERP with PLM, PLM with MRP, and ERP with CRM.

One must also consider the long-term viability of the companies behind the bits and pieces of the enterprise software stack, since small software companies can serve niche industries very well, but a software vendor should be viable and responsive for the next decade or so, with solid customer support, training, and updates as a must. We will put our beloved design data in someone else's hands, so they must be reliable hands. This is also why startups have it so hard to sell enterprise software.

An underestimated factor around enterprise software is the overhead of administering and managing it, and the IT resources it tends to consume, together with the inescapable and eternal *cloud versus on-premises* discussion, and the cybersecurity concerns the IT crew will love to lecture you about. Mind you, hundreds of meetings await you if you're entering the process of selecting this kind of software.

And most importantly, during the selection of enterprise software, one must keep the in-house software engineers distracted so they don't end up proposing to develop

---

[3] Enterprise Resource Planning (ERP) software centralizes and automates a company's core operations, such as finance, procurement, inventory, manufacturing, HR, and supply chain, into a single shared database and workflow environment. Its purpose is to ensure that all departments work with consistent information, reducing manual coordination.

[4] Product Lifecycle Management (PLM) software organizes, tracks, and controls all data, processes, and decisions associated with a product from its initial concept through design, engineering, manufacturing, service, and end-of-life. It provides a single source for product definitions—such as CAD files, requirements, BOMs, change orders, and configuration histories—so that teams can collaborate consistently and maintain traceability across the product's lifecycle.

the enterprise software stack themselves, because they will claim they can make it cheaper and faster, which are both factually untrue. It is always a good idea to send them on holiday when the enterprise software question starts to heat up.

However, lost in the discussion above lurks perhaps the toughest problem when selecting enterprise software: a lax definition and awareness of the product hierarchy. Companies that design systems can go to very long lengths without making the product's hierarchical composition explicit and visible. And engineering enterprise software breathes on it. It must know it, and deeply. There has to be a solid definition from system to subsystem, sub-assembly, component, and part. Moreover, engineering enterprise software needs a unified agreement on how elements are identified, with designators, part numbers, and whatnot. Going shopping for enterprise software with all these details undefined can be highly problematic.

Wherever the road might go in terms of enterprise software selection, it has to be done with an absolute, relentless, inflexible sponsorship from those in power. Adopting these kinds of complex software suites across the entirety of an organization needs industrial loads of buy-in, discipline, and adjustment to new workflows. Any partial approach will keep people wishing to have their spreadsheets back.

And just pray that your company will never merge with another company during your time there. That, right there, is the grandmother of enterprise software shitshows. That's the first thing I think about when I read about mergers or joint ventures. Can you imagine? Two (or more) massive, chaotic, flawed, fetid enterprise software stacks colliding in a beautiful, violent software supernova with a deadly fallout of XML files with random schemas, broken databases, and conflicting naming conventions. Pure hell.

Chapter 16
Ballooning Scopes

Abstract When projects are allowed to have too many moving parts, the solution space becomes too roomy. Indolent oversight and the absence of clear boundaries open the door for continuous addition, with too much flexibility in accepting new requests, incremental changes, and the failure of governance structures to impose discipline on evolving specifications. Change happens, but projects lose their north when scope is let run free, and there are human and organizational factors that allow that to happen.

Keywords Project management · Scope creep · Feature bloat · Featuritis · Incremental complexity · Project anti-patterns · Project management

On August 10th, 1628, the Swedish warship Vasa set sail on her maiden voyage from Stockholm's harbor. She was meant to be the pride of the fleet and a floating symbol of power. She sailed for slightly more than a kilometer before a mild gust of wind tipped her over. Water poured through the open gun ports, and she sank before the crowd had even finished clapping.

The Vasa is one of history's best examples of what I mean in this book by "broken engineering". But why exactly?

The ship's specifications were changed mid-construction by King Gustavus Adolphus himself, who wanted "more guns, higher decks, and more sculptures". Each new royal whim was obeyed by shipwrights who were too afraid to say no. The engineers, trapped between the laws of physics and workplace drama, submitted. They added the extra stuff so the King wouldn't be pissed, violating basic principles of stability. Then they ran some half-hearted tests, noted the ship was dangerously heavy, and chose what broken engineers usually choose: to hope for the best. In the meantime, no one wanted to tell the king his ship was growing too fast, no one wanted to delay the schedule, and no one wanted to look like the pessimist in the room.

On the day of her launch, the Vasa unsurprisingly sank. The subsequent investigation found that everyone had seen the problem coming, but no one bothered to stop it.

I. Chechile, *Engineering is Broken*, https://doi.org/10.1007/978-3-032-18169-5_16

Centuries later, the Vasa sits in a museum in Stockholm. I had the chance to see it. It is imponent, beautifully baroque, and stands as a historical monument to the kind of failed project where managerial stubbornness, ill communication, and people not giving a monkey's come together.[1]

Boyle's Law of Project Management states that if scope is allowed to change freely, its rate of change will exceed the rate of progress. Like many observational "laws", it sounds like something purely out of Wikipedia until you have lived it, as you watch a project slowly drown under the weight of its ever-expanding ambitions.

In every engineering project, scope is what you have promised to do, whereas progress is how much of it you have actually done.

In an ideal world, they should live in harmony: scope defining the destination, progress marking the journey. In practice, they drift apart.

Scope is a boundary, a box, a contract with reality. It defines what's in and, crucially, what's NOT. The problem with scope is that it's abstract. It lives in the heads of those in charge, in a loose collection of documents, and it's kept at bay only by those who give a damn to do so and have the authority to calibrate things back when scope is begging to creep.

Progress, on the other hand, is more measurable: bricks laid, lines of code written (or should I say now fetched from your favorite LLM), components inserted in a board.

The problem materializes when the scope expands more freely than progress moves forward. It's like a marathon where the finish line moves faster than the runner. When this happens, the project begins to feel like it's no longer anchored to reality, and instills the idea that it is constantly under construction, and that finishing feels farther away today than it felt yesterday. The good news? Today feels closer than it will feel tomorrow. Carpe diem.

[1] There's now something called the Vasa syndrome, which refers to the need to stay realistic in terms of strategy and project management, and how organizations need to keep their goals matched to their capabilities. You've got to love it when a project goes so wrong that a syndrome is coined in its name.

Chapter 17
Death by a Thousand Pages

Abstract Designing complex systems is not only a matter of solving technical challenges and doing calculations. It's also about swallowing tons of written content in the form of standards, reference manuals, regulations, and overall documentation that can easily span hundreds of thousands of pages. An engineer working on aerospace, medical devices, or autonomous vehicles must simultaneously fight with the intricate physics of the problem at hand and a monstrous body of written text to process. With system complexity growing exponentially, the informational overload will only get worse.

Keywords Standards · Technical documentation · Reference manuals · Documentation overhead

Without a doubt, engineering complex systems requires surfing a tide of documentation in the shape of standards, specifications, and technical manuals. Author Andrew S. Tanenbaum famously said that the good thing about standards is that you have so many to choose from.

Things have evolved in a way that an almost infinite number of pages eagerly await being consumed as we venture into designing technology, be it the infrastructure that manages our mobile networks, the avionics that control our satellites and aircraft, the radars that will monitor our skies, or the software that runs in basically everything.

The biggest problem with standardization is that standardization bodies do not seem to talk to each other. There's a lack of standardization of the standardization.

It is rather common to find overlapping standards, or standards that point to some other standards, only to find another reference there back to the original in a circular, massive bureaucratic loop.

If we are designing data-intensive systems, we most likely need our systems to interoperate with other systems, so we have to choose standardized ways of exchanging data back and forth. Then, as we fill our cart with the most popular standard out there, we soon realize it is a whopping 7000-page-long treat. Granted, no one reads standards from cover to cover, but at least we will have to look closely enough to find specific things we might be looking for.

I. Chechile, *Engineering is Broken*, https://doi.org/10.1007/978-3-032-18169-5_17

Then, quite probably our systems will need to incorporate a breadth of other data interfaces besides the most popular ones. This means you will need to surf between 60 and 70 different specifications in many different revisions and generations.

If our designs are pursuing common mechanical and electrical form factors for industrial systems, standards are again lurking there, whose standards are specified in more than 100 different documents that will specify things like physical dimensions, signal routing, pinouts, connectors, reliability, and signal integrity matters. And specifications will cost you money (one specification I purchased recently cost me 750 dollars for 64 pages). Not cheap, eh.

And if our design process includes developing board-level electronics, there are standards you will have to comply with for manufacturing and regulatory purposes. Standards that will specify things from how you need to draw footprints for surface-mount components in a PCB to how to fold a through-hole resistor lead with a pair of pliers or manually solder a capacitor, or how to name symbols in a schematic diagram. Electronics standards are captured in around 300 different documents.

The growing complexity of the semiconductor devices used in computer-controlled digital systems does not make it simpler. For all System-on-Chips out there, there is a Technical Reference Manual describing the inner workings of the device, with at least a thousand pages. And because these kinds of devices internally host a CPU core that implements a third-party Instruction Set Architecture (ISA), the reference manual for that ISA is surely several thousand pages. A modern, commercial Reference Manual for a popular core is about 13,000 pages long.

Can it get worse? Oh yeah. If our design were related to mobile networks, this would also require a good dose of reading. Credit where it's due: modern mobile technology is pure engineering genius resulting from industrial amounts of consensus gathered in complex global consortia. However, that consensus comes with its page count. By some accounts, there might be more than 100,000 pages worth of specifications by the standardization body of fifth-generation mobile technology.

And because most systems out there include software, there are standards for software as well, specifying things such as operating system binary interfaces, web browser data types, and whatnot. Last but not least, any "connected" system will have to comply with the necessary Internet protocols, and this would require getting acquainted with the Internet Engineering Task Force (IETF) and its more than 9000 technical standards with diagrams drawn in ASCII. Again, not every standard and document applies to our design, but still.

Needless to say, this is far from being a critique of standards as a concept; standardization is the invisible fabric that makes modern life possible. Those cryptic ASCII diagrams buried in IETF documents are why you can wirelessly stream Taylor Swift at the gym without thinking twice about it. Bluetooth connects your phone to your earphones using IEEE 802.15.1 protocols. Wi-Fi (IEEE 802.11) pulls the music from Spotify's servers using HTTP/2 over TCP/IP. AAC or MP3 standards decode the audio file. Your phone's ARM processor follows instruction set standards, while the app itself relies on countless software APIs and data formats. Every wireless frequency is internationally coordinated to prevent interference. Every packet of data follows agreed-upon structures. Every cryptographic handshake securing your connection implements standardized algorithms.

Conversely, lack of standards creates chaos. First and foremost, the chaos of not being able to listen to Taylor's music. Think of when railroad companies used different track gauges, forcing passengers to change trains at borders. Think of the current mess with electric vehicles' charging connectors, where drivers must find compatible chargers or carry adapters.

But standards are definitely getting too complex, coming a step closer to the point where they might become unusable. A rough estimation tells me there might be more than one million pages of documents between specifications, manuals, and handbooks waiting for us when designing interoperable systems of reasonably complexity. A typical A4 page may contain around 450 words, so we're talking about 450 million words. We as adults can read at an average of 250 words per minute, which means that we would need 30,000 h to consume it all. It means roughly 3.5 years, nonstop, without even sleeping. Considering we typically work 40 h per week, it would take 14 years of our career, where the only thing we would do is read standards and manuals, full-time.

At any rate, this can only get worse. Considering the increasing complexity in every aspect of the systems we design, it might very well be that a few years down the road, we might be talking about 10 million pages of standards, manuals, and the like.

It's not so wild to think we are heading to some "crisis of documentation", where standards and technical documentation will have to be rethought, where technical insight will need to be conveyed in some other format other than the "book" style. The book style is simply not scaling.

17.1 The Interoperability Paradox

I spoke about standards and the benefits of interoperability in the section above; if interoperability is so beneficial, why aren't our systems more interoperable? Do we care about interoperability as engineers?

Picture a parallel Universe where any data produced by any system can be directly understood, processed, and augmented by any other system that has complementary capabilities the former system does not have. This hypothetical scenario belongs to an almost unthinkable Universe where all human-made systems are capable of talking to each other, where interoperability and collaboration are the norm, by design. A Universe that could be (or could have been), but it is not and perhaps will never be. It appears it is in our best interest to continue creating siloed systems.

This is the fundamental challenge with standards: achieving total interoperability would require finding a sweet spot that satisfies countless competing interests simultaneously. First, there are protection concerns that come in different flavors—data protection, intellectual property protection, privacy protection, and access protection—largely driven by competition and regulations. It's worth noting this isn't synonymous with encapsulation, as we can have a properly encapsulated system perfectly capable of interoperating. Beyond protection, cost considerations create a powerful disincentive, since it appears cheaper and faster to do things without caring about interfacing with others. Market dynamics further complicate matters, as system makers naturally resist giving competitors the chance to interface with their products and potentially profit or extract proprietary information in the process. Perhaps most fundamentally, achieving consensus requires manufacturers to sit down and hash out implementation details, and inviting other organizations to join means opening up conversations and disclosing substantial information, a process that, while being the very reason standards exist, demands insane amounts of work to accomplish. And takes time, which delays time-to-market.

Not surprisingly, the lack of interoperability in the systems we make is the result of our intrinsic self-interest. And our selfish behavior does not only show when talking with external entities; it also shows internally. We also design and grow siloed systems within the same organization. Internal silos emerge from the same forces that create external ones. Teams optimize locally because it's faster; why spend weeks coordinating with another department when you can ship your feature today with your own custom solution? Each team becomes protective of its domain, its interfaces. "We can't change that interface; it would break everything" becomes the common reaction, even when the change would benefit the organization overall.

17.2 Thirty-Thousand Requirements to Rule Them All

It's always interesting to see how certain regulatory frameworks tend to degenerate into something that contradicts their original intent after morphing into a compliance hairball that suffocates the innovation it was meant to (supposedly) protect, and in the process, trying to convince us that overregulation is the best that can happen in a well-functioning society.

This is my story working with one of the most bloated bodies of standardization that humanity has ever conceived. This particular one comprises nearly thirty thousand separate requirements distributed across over a hundred active documents. In case you are wondering, yes, all that on top of all the documentation and specifications that naturally originate in any technical project.

I will talk in the present tense because this travesty still exists, although it is (luckily) in my rear mirror, and I hope I will never come across it ever again.

First off, this bureaucracy behemoth starts off by funnily claiming to be prescribing the WHAT but not the HOW, whereas in reality, all their energy goes into making sure the HOW is done exactly as they say, or else. That's a promising start. In this "initiative"—as they like to call it—the distinction between prescribing the system being engineered and prescribing the process for engineering it becomes idiotically blurred, with mandates targeting both without clear separation. From the perspective of these standards, your and your colleagues' education, experience, and lessons learned as engineers are meaningless and duly disregarded: the system suggests methodologies and approaches at every turn, reinventing a thousand wheels. All while confusing the subject of engineering work left and right; at times it targets the system under design, and some other times it targets the process that devises the system. This is a common occurrence these days in our craft, as if engineering schools had forgotten teaching the difference between the two.

Although this gigantic bag of documents is self-perceived as "standards", what you find is fundamentally different from true standardization. As we discussed in previous sections, a key characteristic of real, useful standards is that they ensure interoperability, or safety. What we have here instead is a disjointed, sloppy collection of documents without interoperability in mind, in a way that you could have two systems that fully "comply" with this disgrace and still be completely incompatible with each other.

Additionally, there's no formal certification or appraisal process; a distinctive feature in otherwise serious standards, where you must demonstrate objective compliance to receive recognition, so others can rest assured that they can interwork with you. In fact, the entity controlling these "standards" is adding disclaimers everywhere that remove any responsibility from them in the use of these standards. Sussy.

Some might say that the thirty thousand requirements do not apply to every project, and that is true. But believe me, the process to understand which ones apply to your project versus which ones don't is as excruciating and experimental as can be.

An interesting fact is that this system of standards is enforced largely through contractual coercion. Because there is no official verification that you have truly met the requirements, the burden of demonstrating compliance remains ad hoc, executed by a set of enforcers who are committed to the cause, although poorly trained in the "system" they are supposed to be enforcing. These enforcers, unable to verify thousands of individual requirements by any practical means, zoom out and focus instead on a handful of bureaucratic artifacts. In the way this "thing" is conceived, every requirement can vomit a document; therefore, these regurgitated documents become the currency of project approval and payment milestones.

It is fair to say, not every single document in this degeneracy is pointless. Because this system of documents is arranged hierarchically in a tree-like structure, it has distant branches where some selected standards are, in fact, useful and can guide the design process in the right direction, a direction of slightly higher safety and interoperability. However, we are talking about the most specialized standards in all branches, which might account for maybe 1% of the whole body at the most. All the rest is, well, noise, and it could disappear literally tomorrow that nothing would happen. This whole thing could be in fact 300 requirements (which still sounds like a lot).

Every technical project in the history of humankind intrinsically generates information during the development stage in the form of analysis files, test reports, design diagrams, and spreadsheets. This is the productive messiness of creative work, bringing order from disorder. And yet this standard system keeps on attention-whoring by forcing you to structure and present this information in the way they want, or else. They even ask you to produce "justification documents" that require you to demonstrate how your design meets the requirements. Isn't that literally the entire point of the development process? Don't we justify our work by iterating, testing, verifying, correcting, and refining until it works? Can a single document truly capture the quasi-infinite sequence of decisions that constitute justifying a complex problem-solution fit?

Few professional experiences have felt as exhausting for me as navigating the compliance labyrinth related to a body of standards like this. The sad part, though, is when you see engineering organizations boasting about their compliance with this infamy—even when not required to!—just for legitimacy. It is incomprehensible to me how the minds of certain people work to consider voluntarily adding all this noise to their lives. But hey, masochism can take many forms: some like to be burned with wax or hit with a stick, some others have a kink for compliance with bloated standards that add no value. To each their own.

The unnerving part is that a lot gets lost in this nonsense. Time that could be spent solving actual technical problems. Energy that could drive real innovation. Talented people who could be using their gray matter for building real technology instead of spending their days trying to please the occasional bureaucrat. And perhaps worst of all, all this creates a perverse mechanism that favors those who are good at playing the red tape game over those who are good at actual engineering. The organizations that thrive through this painful process are not necessarily the most innovative or technically capable, but those with the resources to feed the bureaucratic monster. All while small teams with good ideas but limited patience and resources give up or go elsewhere, taking their ideas with them. If anyone still wonders why progress feels slow, or why the most exciting work seems to be happening somewhere else, the answer is right there, buried under thirty thousand requirements.[1]

---

[1] This section was once an article I published in a blog that I used to maintain. The article sparked an interesting discussion, with people who were aligned with my perspective and some others who were passionately against it. Who were the ones angered by my words? Of course it was the bureaucrats. If you are reading these words, no hard feelings! Keep reaching for that rainbow!

Chapter 18
Engineering Myopia

Abstract While the notion of "seeing the big picture" is often praised in engineering circles, it can easily transform into technical myopia. The integrity of complex projects depends not only on high-level vision but also on how thoroughly their interstices are explored. Coverage is what determines whether the full range of a project or a design has been observed, understood, and acted upon. Just as a lazy pianist may ignore half the keyboard, lazy engineers often selectively focus on familiar "hotspots" while neglecting remote, less interesting areas. Managing complexity requires moving fluidly between macro vision and microscopic attention to details, where the devils love to lurk.

Keywords Big picture · Project coverage · Complexity · Project management · Systems thinking · Project coverage

Everybody likes to preach about the benefits of seeing "the big picture". But what does that mean? It appears that seeing the 'big picture' refers to the ability to understand a situation or concept as a whole, rather than getting bogged down in specific details. The concept of the 'big picture' is better explained by the idea of coverage.

In the software industry, the term 'coverage' is more frequently found and used for referring to a percentage measure of the degree to which the source code of a program is executed when a certain test suite is executed. A program with high test coverage has more of its source code executed during testing, which suggests it has a lower chance of containing surprises—i.e., bugs—compared to a program with low test coverage.

Now, let's change the landscape. Picture yourself playing piano, more specifically a piano with 7 octaves, like most pianos have. If you are a beginner or mid-level piano player, you will mostly stay in 2 or 3 octaves at the most, so you are using half of the keys the instrument offers. The rest of the keys are seldom pressed by us mortals.[1]

[1] The pianist most famously known for exceptional reach and wide keyboard use is Sergei Rachmaninoff, whose massive hands allowed him to span up to 12 or 13 keys (an octave plus a major or minor sixth), enabling his famously dramatic and difficult chords. You wouldn't like to get slapped by Sergei.

I. Chechile, *Engineering is Broken*, https://doi.org/10.1007/978-3-032-18169-5_18

When it comes to executing a project that is composed of a blend of many parts, activities and sub-activities, coverage refers to the somewhat unquantifiable measure of the degree to which all the relevant parts have been explored and observed by the relevant actors. Complex projects and systems describe a territory or a map of sorts, with its dimensions and boundaries defined by all the elements, components, and tasks necessary for its completion. In most projects, engineering managers do not tour said territory to its full extent but stay mostly around certain centers or 'hotspots', ignoring the outskirts and the tough neighborhoods. It is quite typical for managers to stay around parts of the territory that are more comfortable to them, more related to what they know and what they like, or what their fingers can reach. We don't usually find *Rachmaninoffs* in engineering management, but managers with tiny figurative hands.

18.1 Seeing the Crumbs

I have quite bad eyesight. I use two pairs of glasses, like older people do: one for looking at distant things, one for looking at nearby objects. I change them all the time as the day goes. As annoying as it can be, it's so much a part of my daily flow that I just don't notice the nuisance anymore.

Among several disorders, I have hypermetropia, which is a condition in which I can't see objects near me. When dinner is over at home, and I start to clean up the table, it's always interesting to see how different the table looks to me with and without my glasses. I like to do this silly exercise where I clean without glasses, only to then put on my reading glasses and see how filthy the table still is. The same happens when I look at the sky with and without my other glasses; I can see many stars disappearing as soon as I take them off; the sky suddenly becomes unsettlingly dark, except for the brightest ones (and the Moon, which luckily, I can still see even without glasses).

I learned eons ago not to trust life around me without glasses, for my eye condition might be hiding important details of my surroundings. I also heard from people who got eye surgery that one of the main drawbacks is that you become instantly aware of all the dust around.

Standing on this analogy, engineering projects are like my table after dinner: full of crumbs. And, as a manager, you either see the small stuff or you don't, depending on what glasses you put on.

With all, the so-called "big picture mindset" is equivalent to confessing you have a raging hypermetropia. You choose not to see the crumbs. The question is whether you have a pair of reading glasses at hand if needed, or if you choose to go about your professional career seeing life as a constantly blurred image.

Projects come to fruition when absolutely all the things, including the tiniest of the nuances, the minutiae, and the incredibly specialized small stuff, come together. There's no point in building a house without ensuring the toilets have their floats or assembling a bicycle without the bolts that keep the brakes attached to the frame.

Things—systems, products, toilets—become useful and successful if everything, accounting up to the last bit, converges in a more or less harmonious way.

In general, having a "big picture" mindset can be useful. It may help you abstract yourself from the buzzing details, observe things from a more abstract distance, and gain knowledge about a situation. But you cannot go on in life the myopic way. You need to focus when needed.

There are too many myopic engineering managers out there, with no glasses at hand, leading things and practicing the "big picture" religion. In fact, these "big picture" folks are the first to be shocked, and at times annoyed, when exposed to the fact that the table is full of crumbs, that the project is not going anywhere because all the minutiae that they neglected for such a long time have now become relevant and are blocking everything.

Chapter 19
Appetite for Destruction

Abstract The engineering practice suffers from the effect of changing for the sake of changing. It's easy to ignore why things are the way they are or how much sweat, sacrifice, and work hours something took to build, only for the occasional reformist to dismiss it as "a mess" that needs to be changed completely. The asymmetry between construction and destruction is too underrated. Building requires creating order from chaos, lowering entropy. It demands energy, time, patience, and effort. Destruction is practically free. Every mess has a story behind it, a context, multiple constraints, and trade-offs made by people who had their good reasons, only that they were operating under different circumstances.

Keywords Change management · Entropy · Construction · Destruction · Improvement · Engineering process · Change control

Oscar Wilde wrote:

> Yet each man kills the thing he loves,
> Some do it with a bitter look,
> Some with a flattering word,
> The coward does it with a kiss,
> The brave man with a sword!

Destroying the things we love is not some relic of humanity's dark past. Human sacrifice was practiced by countless ancient cultures, with people and animals ritually killed to please or appease gods and spirits. Droughts, earthquakes, and volcanic eruptions were interpreted as divine ire, and blood offerings were meant to calm the celestial temper tantrum.

We have evolved, supposedly. We don't slaughter goats on altars after storms anymore. However, in our modern engineering organizations, we still put our figurative sacrificial goats on the table. We still kill the things we love. We destroy the things we built with passion and patience.

The asymmetry between construction and destruction has always been somewhat underappreciated. Building requires creating order from chaos, lowering entropy. It

I. Chechile, *Engineering is Broken*, https://doi.org/10.1007/978-3-032-18169-5_19

demands energy, time, patience, and effort. Destruction, by contrast, is practically free. It produces a quasi-random, useless outcome with a single blow.

A craftsman needs weeks to build a guitar; smashing it to pieces takes two minutes. An adult goat takes fifteen years to mature; slitting its throat to appease an imaginary deity takes seconds. Forming a genuinely jelled team can take a decade of careful cultivation; dismembering it in the name of half-baked metrics takes one PowerPoint presentation between two clueless managers in a distant conference room who have never met the people they're about to upset forever.

Perhaps the most perplexing self-destructive act occurs when the perpetrator has been intimately involved in building the very thing they're destroying. Having a first-hand understanding of how much effort and energy something required, knowing the intricacies of its mechanisms, the subtle complexities, the pains endured, the obstacles overcome; all of this should weigh heavily on the executioner's mind while wielding the ax. So why doesn't it?

Some people will sell their own mother if their ego is massaged the right way. Offer them a position of power and a shinier title but first ask them to burn with a flamethrower all the work they have done over the past decade as a loyalty oath to their new "deity" (some boss), and many will do it without hesitation or remorse. The dopamine bomb of achieving hierarchical status causes amnesia in many people.

There's also the pervasive "illusion of improvement" that invades the minds of newly minted managers. Drunk on their newfound authority, they're simply destroying what took years to put on its feet. It's always amazing to see how power stupefies otherwise intelligent people and atrophies their judgment. Many great engineers make incredibly stupid managers. A good soldier in the trenches can make a terrible general in the war room; being competent at tactics doesn't qualify you for strategy. And by the time they realize they have destroyed something that once functioned, they're too far removed to get the shrapnel, perhaps one or two promotions away, safely protected from having to clean up the mess. And the current way of doing things seems to reward mindless change.

There is this story going around the Internet of an engineer at a FAANG who had inherited a legacy software service that had been running flawlessly for five years. The system had zero issues and required minimal maintenance. It was the kind of boring, reliable stuff that companies depend on (but seldom celebrate). For an engineer looking to advance to his next hierarchical level, this was a problem: there was nothing broken to fix, no crisis to solve, no opportunity to demonstrate the kind of complexity that bosses seem to love and reward.

Thus, the engineer proceeded to write a 20-page design document arguing that the system needed to be "future-proofed" by rewriting it entirely in another language with a different architecture. The actual system didn't need this transformation, as it was working perfectly. But something had to be done; its stability was not creating enough visibility for him. The proposal framed the current, reliable service as "legacy" and how it was holding the organization back, and positioned the total rewrite as a modernizing initiative to prepare for what was coming, without ever specifying what was actually coming.

The project consumed a full year of engineering time. Three other engineers were pulled into the charade. When the new system finally launched, it was worse than what it replaced. The rewritten service was slower than the original. Infrastructure costs increased. However, because the launch materials were crafted carefully, filled with bombastic language about modernization and continuous improvements, the engineer received their promotion shortly after.

The story could be yet another fake story on the Internet (is there anything real these days?). However, this is really how "the system" works in many places. When companies define their promotion pipeline as a function of complexity and mindless future-proofing rather than stability and reliability (two fundamental attributes of good engineering), they get exactly what they want, and that is engineers who build unnecessary things and leave the mess behind by climbing up the ladder before the feces collide with the ventilation device at terminal velocity. This also tends to be called Resume-Driven Development, or RDD.

It's easy to be a reformist. For the reformist soul, everything done thus far is garbage and must be replaced with something better, more efficient, more modern.

The usual behavior for newly-appointed managers is: "I'm here now, things will be very different. I represent tidiness, and rational thought. The past represents the exact opposite; chaos, and inefficiency,".

However, hiding beneath the reformist's mind lies a combination of laziness and a thirst for dominance.

In nature, when a new coalition of males takes over a lion pride, they almost always kill the cubs. They're not biologically related, and the new lions have no interest in spending energy ensuring another male's genes get passed on.

The corporate reformist operates on similar instincts, though less gruesome. They need to kill the past because the past cannot be allowed to persist and propagate. They represent a fresh start, a new DNA.

It's easy to opportunistically overlook why things are the way they are or how much blood, sweat, and work it took to build what the reformist dismisses as "a mess". Every mess has a story behind it, context, constraints, and trade-offs made by people who weren't idiots, but were operating under different circumstances.

Sure, improving real messes is necessary. However, every change creates new messes. If you're convinced you have the perfect solution for something, understand that someone in the future will inevitably flag your solution as a mess that needs fixing. Then someone will do the same to them. And the construction-destruction cycle shall repeat.

New bosses will arrive with their preferred toolkit—the methodologies, frameworks, and buzzwords they have always used—and will demand everyone switch unconditionally. This only shows unwillingness to learn and adapt to the existing order of things, but also arrogance, refusal to adjust to different landscapes and intellectual laziness to ride new learning curves.

Everything can technically be labeled a mess because nothing is perfect. But does everything need to be burned down and rebuilt from scratch? Not really.

Reformist souls should always remember that the problems of today were meant to be the solutions of yesterday, and that someone worked hard to build what you're about to destroy.

Chapter 20
The Cold War Between Sales and Engineering

Abstract Sales and engineering fight all the time, although they need each other badly. They are nothing without the other half. The best product in the world will not sell if there isn't someone with the expertise to deeply connect with potential customers and instill in them the fear of missing out. Equivalently, the most charismatic sales team ever will starve to death if there is no real technology to back the grand promises. Will sales and engineering ever live in peace? Which side does Product Management take in this everlasting feud?

Keywords Product management · Technical sales · Product customization · Product-market fit · Business development

Ask an engineer, and they will say salespeople are simpletons who promise the world without having a clue how to deliver.

Ask a salesperson, and they will say engineers are emotionally challenged narcissists who cannot understand the slightest nuances of the world around them.

The act of developing complex technical products is radically different from the act of developing a business for those products. And in that difference lies one of the longest feuds in any organization selling technology. I haven't had the pleasure of working in any organization where you couldn't feel the tension in the air between sales and engineering. Imagine that you visit a car dealer to buy a German sedan. You would never think of asking: "Does this car have six wheels?" When the salesperson says no, you wouldn't respond with: "Can you ask the engineers back in Germany if they could add the extra wheels? Tell me how long it would take and how much it would cost".

This doesn't compute. The car is like it is. Although we do live in an era of high customization, where you are in fact able to ask for some selected changes in the car's features, you can't ask for random changes that are not included in the official product variants. There's a wide air gap between you, the car dealer, and the car's R&D team back in Germany. Once the brand-new car is out of the factory and loaded in a lorry, off it goes. Take it or leave it. There is nothing the car salesperson can do to accommodate your eccentric request for extra hardware.

I. Chechile, *Engineering is Broken*, https://doi.org/10.1007/978-3-032-18169-5_20

However, when you narrow the geographical gap between sales and the R&D workforce, and they happen to sit under the same roof, that's a different story. And because certain engineering projects seem to radiate this aura of "the sky is the limit", that is precisely when the mouse and the cat start chasing each other, when salespeople promise what's not there ("the number of wheels is configurable"), and engineers get angry because adding it would imply a lot of pain and tears and basically change the whole system. Or also—and to be fair with salespeople—when engineers are reluctant to develop features that the market is clearly signaling are needed. There is no right or wrong in this. Both have valid points on their own:

- A good engineering team must adapt to the evolving needs of markets
- A sound sales team must filter some signal from the noise; not every wild request can reach the far ends of the engineering dungeons

The feud gets worse with desperation. When the business is languishing—and considering that salespeople tend to work on commissions and bonuses—this may create a fertile soil for promising the world and beyond.

Selling technical products requires good technical knowledge of it. However, it also requires soft skills, such as good communication, charisma, and a sense of business opportunity. When you only go hardcore on the technical but lack the necessary salespeople's skills, you can only expect to sell to customers who are also very technical and are prone to fall for your technical bravado, which are rare. There are too many companies out there purely led by engineers, making no business at all, flexing on technical prowess and intricacy as their sole marketing strategy.

Equivalently, a company wanting to sell technical things with only salespeople and no engineers would never pass beyond vaporware.

Sales and engineering need to coordinate and collaborate in order to exchange the knowledge necessary to keep the sales process self-sufficient, and bring the techies in to sit around the table should the discussion really need it. Endless customization sounds great, but it's terribly expensive and demanding. Is there any better way to couple sales and engineering together? Something that would handle the constant tension?

That's when the (still) somewhat underrated idea of Product Management comes into play, acting as the clutch, the funnel and mediator between dogs and cats, the deuteragonist (what a word, I love it) in this story, just like Spike in Tom&Jerry. We expand on this in the next section.

All in all, it's important that engineers understand that we are practically always working for a product and that the product needs to fit into a market.

20.1 Product Management to the Rescue

I briefly mentioned Product Management as the clutch between Sales and Engineering, and this begs for some elaboration. A brief mea culpa first; I didn't realize the importance of Product Management until maybe a decade ago. Before that, I had a bit of the typical macho engineer attitude when I thought everything could be handled by self-sufficient engineers working together in a brotherhood of technical awesomeness and unstoppable featuritis. Until it became clear to me that engineers alone don't get too far.

Proper, serious Product Management is a fundamental organizational capability that takes ownership—real ownership, not the vapid kind we spoke before, and from start to finish—of the life cycle of anything you intend to sell. Your technical object can be tangible or intangible, small or large. But if you're going to sell it and someone's going to buy it, then it's a product, and it deserves to be treated as such. With respect.

That means investing time and resources to understand who needs it, what problems it solves, what its strengths are (so you can amplify them), what its weaknesses are (so you can put some make-up on them), and what alliances you'll need along the way with suppliers, tools, partners, and so on. That's product management in a nutshell, which also helps in building product families, where you can reuse common elements across the product line, lowering costs or at least saving time by not reinventing the wheel time and again across your portfolio. Product Management helps tame the combinatorial mess of product variants, classes and generations.

Solid product management also creates good sales material, strong brochures, curated technical content, all stuff which is essential in selling.

Good product management shows. Mediocre or nonexistent product management shows too.

When I talk about product management, I'm not referring to the innocuous focus groups and the empty market research propelled by management consultancies. I mean something grounded and direct. You can do product management without falling into empty corporate wish-wash. Go out, meet potential or current clients, talk to them, squeeze every last bit of information out of them, process it, distill it, share it internally, and then gather the relevant people around the table to design and build it so the odds of success go up. Who could be against that?

Well, Product Management meets a lot of resistance from people like me a decade ago, mainly because good product management takes time. And to do it seriously, you need to educate minds along the way. There's still a ton of ignorance around what product management really means, and plenty of products out there born from gut feelings or impulsive crusades by organizational emperors in chief who assure you that "this is the way" but can't show a single data point to back it up. There are plenty of product managers hired only to be ignored, left to chase decision-makers until they finally give up and move on to a place that takes the craft seriously. And so we march into design and development processes justified by guesswork, by engineers' gut feeling, taking years to produce products that are most likely obsolete by the time they see the light of the day. There's still too much attachment to "the solution", while barely anyone spends time truly understanding the problem.

A long time ago, I had an endless argument with my then-boss. We were debating whether products are born "from the inside out" (meaning the company dreams up an idea and then tries to sell it) or the other way around, where you "listen to the market" and build from there. He was in favor of the first. I was in favor of the second. We didn't agree. It happens.

When you go from ideas to market, the problem is that you usually go overboard. You develop at full throttle, open-loop, pure adrenaline, eager to launch as soon as possible, without really understanding what's needed. When you finally launch, you crash straight into a wall and discover all the problems at once. In this romantic, idea-driven approach, it's perfectly possible to realize halfway through that the product was unfeasible or too complex for your capabilities, or that there was actually no problem to solve.[1] By then, you're in too deep.

The other approach requires a thorough analysis of what's actually needed. From there, you develop the product that fits the need, but only once you truly understand it. With data, not opinions. This way, you hit the problems early, and by the time you emerge from the design labyrinth and are ready to go to market, the product is more refined, aligned with the signals your analysis gave you, and relatively free of nasty surprises.

Sure, you can't just sit and analyze forever or drown in a sea of metrics; you will starve to death if you do so. You've got to move; sketch, prototype, test, iterate (with clients in the loop), and keep going. There will always be surprises, of course, but it's better that they show up early; everyone knows that the later you change something in a design process, the more it costs.

That's what good product management looks like, and it deserves time, attention, and resources, without ever losing that crucial, healthy sense of urgency. Because apparently, there's never time to do things right, but always time to do them twice, although in many cases, there's neither time nor money left to try again.

[1] Juicero (2013–2017) exemplifies the problem of *solutioneering*. The company raised $120 million to develop a WiFi-connected juicer that squeezed proprietary juice packets, until reporters discovered the packets could be squeezed…by hand. The device essentially automated what didn't need automation, solving a problem that didn't exist while creating new ones: vendor lock-in, unnecessary complexity, and a price point that made fresh juice prohibitively expensive. The company shut down in 2017.

Chapter 21
Project, Systems and Products

Abstract The distinction between projects, systems, and products remains surprisingly inconsistent in a typical day in the engineering workplace. Each represents a different temporal and conceptual relationship to engineering work, and understanding their differences matters, not only for semantic correctness, but also in terms of how organizations allocate resources and manage expectations for each one of them.

Keywords Project management · Systems engineering · Product management · Products · Systems · Projects · Terminology · Disambiguation

Ok, let's go right to the definitions. We start with the project. A project exists to achieve a specific objective and then conclude. When an organization undertakes a project, it commits resources toward a defined outcome with an anticipated endpoint and is bounded by time and scope. Yes, the problem is that the endpoint often shifts consistently to the right, but in theory, it shouldn't constantly move. Projects are instruments of change, whether that means building something new, improving something existing, or solving a pressing problem. Projects consume budget, demand coordination, and ideally deliver something tangible before disbanding the team and releasing the resources. The project's essential characteristic is its temporality. It is meant to end, and the fact that we struggle so much to end them is a good reason why we tend to get confused, because certain projects seem to have been there forever, so they gain themselves an entity of their own. Be sure that many engineering organizations make a living out of projects alone; they are not product-oriented, they do not have an established product line and their business model is basically to monetize their capabilities to create different complex artifacts needed by customers.

Systems are intentionally (and hopefully) enduring. When talking about systems, it encompasses not only the technology itself but also the activities, the knowledge, and the maintenance and operations routines. Systems are what projects leave behind, or what existed before projects came along to modify them. They require maintenance, periodic updates, and eventual replacement. They are designed to persist, to be reliable, and in cases to fade into the background of routine operations. The

I. Chechile, *Engineering is Broken*, https://doi.org/10.1007/978-3-032-18169-5_21

system's essential characteristic is its continuity for as long as its design life span (statistically) stipulates, although it may suffer premature mortality.

Products occupy a different category entirely. A product is an offering; something created with the explicit intention of serving users or customers, whether internal or external, for profit or not. While a project might build a product, and a system might become a product, the product itself is defined by its relationship to a market or user base. Products have competitive landscapes and value propositions. They must be marketed, supported, and evolved in response to user needs and competitive pressures.

Every technical object is a system, but not every system is a product.

The confusion among these terms often stems from their natural progression. A project coordinates the tasks to build something in terms of time and budget. That something becomes a system or part of a system, and if conditions are met (there's a market), it might also become a product. Treating a system like a project suggests it will someday be finished in a way that attention will be entirely switched away. Treating a product like a mere system neglects the demands of the market and all the supporting activities necessary to meet customers' desires.

That being said, projects never exist for the sake of existing. To build a new bridge for connecting two neighboring cities, you organize the work to achieve it, and it's clear all along that what you want to get is a damn bridge. People cannot go visit their relatives across the river on board a Gantt chart or piggybacking a project manager.[1] Every engineering project requires managing interdependent and often competing technical decisions, trade-offs, and discussions needed for developing the outcome, and when project managers are given that authority, chaos generally ensues. That's the territory of Systems Engineers. Product Managers have little to do there; they are supposed to be helping engineers get along with sales so what's being built is aligned to what is needed. As you can see, we engineers love having many open fronts and need a lot of help to get along with others.

The feud between Project Management and Systems Engineering shouldn't really be, since it's rather simple to tell apart which does what, but somehow the usual intellectualizers of work have made it bizarrely more complicated than it is.

Systems Engineering was naturally conceived to take the role of bringing technical substance to the planning by plowing forward with the activities to create the tangible outcome, and in the process, filling any other technical gaps that may (and will) appear in the journey. Project managers, as good neighbors, husbands, wives, sons, and daughters as they surely are, cannot define the technical path or set forth the discussions needed to create complex systems. A technical project without Systems Engineering is like a cargo ship adrift in an ocean of forged requirements and half-baked work breakdowns, feeding unrealistic Gantt charts that are treated as the main outcome of the activity. Run—and run fast—from a project when you feel the Gantt is treated as an end and not as a simple planning tool, which it is.

[1] You could, technically, build a bridge made of many Project Managers lined up, if the water is shallow enough.

However, to be fair with project managers, a project populated only with systems engineers is a project equally destined to failure: baroque, bizarrely overengineered, expensive, and under perpetual featuritis. Using the bridge analogy again, a bridge resulting purely from systems engineers working together will have 20 lanes on each side stacked in a 3D layout, all while generating its own power by an IoT-controlled nuclear reactor. Systems Engineers think time is an infinite resource, that money grows on trees, and that one single feature they so badly want to add and nobody asked for won't create much disruption, when in fact it has the potential to delay the project for a full year.

Project Management and Systems Engineering need each other because they balance each other. Their relationship is a symbiosis[2] rather than an overlapping one. They are the legs any technical endeavor needs for walking from ideation to delivery. Leave an engineering project without one or another and just sit to observe it stumble like a drunken sailor. And if you want to condemn it in the worst way possible, bring them close together without clearly stating who does what, and watch it burn.

[2] Note that symbiosis can also be parasitic, for instance, where one actor just exploits and harms the other one. And yes, some engineering projects may show this kind of dynamics.

Chapter 22
Dangerous Originality

Abstract Engineers are constrained by the context to be somewhat unoriginal, to follow established architectures that have formed with time, iterations, and are shaped by factors they cannot control. This phenomenon, where products within a category look similar across different vendors or manufacturers, is influenced by practical, ergonomic, and regulatory considerations. This sounds as if engineers were "cursed" to be unoriginal in order to succeed. Or worse, that innovation could be a problem. We will see that it's not the case, as long as one is aware of the "curse".

Keywords Design · Imitation · Design constraints · Proven architectures

Given enough iterations, the system architecture for solving a particular problem converges to the most optimal one, if it exists.

The problem arises when engineers ignore this and try to be too original, ignoring that an optimal solution for the problem might have been found ages ago.

In simpler words, if you want to engineer something successful, you may need to stick to creating something similar to what exists. This sounds pretty terrible from the outset, as if we're cursed to be unoriginal in order to succeed, or worse, that innovation could in fact be counterproductive. We will see that it's not the case, as long as you're aware of the curse.

See excavators, for example. The ones used for digging holes in roads and that. They all look practically the same. They all have the obligatory cabin, the articulated boom, the bucket, wheels, or endless tracks. If you were a founder creating the next killer excavator company tomorrow, we could easily predict what your excavators would look like with reasonably high precision.

Now, see cars, for example. They look more or less all the same. Sure, a Prius and a McLaren Speedtail have dramatically different characteristics, but the main architectural traits are the same: four wheels, a cockpit, a trunk, a steering wheel, rear lights, front lights, doors on the sides; you get the score. You could get out of your Corolla and go drive a Lambo without too many problems. Remember, this is about architecture, not about performance or aesthetics. Performance and aesthetics are the factors automakers use to differentiate themselves from each other when they

I. Chechile, *Engineering is Broken*, https://doi.org/10.1007/978-3-032-18169-5_22

are cursed to share the same established architecture that's been around for more than a century.

Now, see aircraft, in particular airliners: the tube-and-wing architecture settled several decades ago as it proved to be the most efficient system architecture. Check any flight over your head, and you will see the curse of unoriginality in motion: the long tube, the two engines under each wing, and the horizontal stabilizer.

This phenomenon, where products within a category look similar across different vendors or manufacturers, is influenced by several factors, often shaped by practical, ergonomic, and regulatory considerations rather than any natural "law".

For starters, certain design features are simply dictated by the function and purpose of the product. We discussed excavators before: well, they surely need a stable base, a powerful arm, and a cabin for the operator. This leads to a similar form across different manufacturers. Digging dirt hasn't really changed as an activity in a long time; therefore, the machine for doing it shouldn't either.

Ergonomics plays a significant role in product design as well. Devices like smartphones share similar shapes and sizes because they need to be held comfortably and fit in our pockets. Go try to convince investors about your great idea of a triangular phone,[1] but you would probably come back empty-handed; there's a reason why they're flat and compact. Or try to design a car that will require customers to re-learn from scratch how to drive. Bombers from the Cold War Era? They all look the same. Sure, a big portion of it was thanks to the tireless work of industrial spies, but laws of physics were the same regardless of which side of the Iron Curtain you were on.

The materials and manufacturing processes used can lead to similarities in design. Certain shapes and structures are better suited to withstand stresses and strains. Also, standard manufacturing techniques (like casting, molding, and machining) may favor certain shapes and designs over others.

Also, regulatory bodies often set standards that products must adhere to for safety, interoperability, and performance. In the automotive industry, vehicles must meet safety regulations that influence their design, leading to similarities. Satellites must meet specific requirements and constraints for fitting in increasingly standardized launch vehicles, which might favor their "box" shape with many stowable appendages due to volume constraints. For commercial aircraft, anything other than the tube-and-wing design would face big challenges while being serviced in today's airports globally, making the design economically impractical, as "cutting edge" as it may appear.

Furthermore, companies often follow trends to meet consumer expectations and preferences. While functional, products also need to be visually appealing and consistent with brand identity. Manufacturers often emulate successful designs to remain "current" and competitive.

Convergent evolution plays a role as well. Originally a biological concept, it refers to the process by which unrelated organisms independently evolve similar traits as a result of having to adapt to similar environments or ecological niches. In design,

[1] Except, of course, if we're talking about Sabre's Pyramid Tablet. The strongest shape ever constructed. Unleash the power of the pyramid.

this concept is used to explain why different designers might independently create similar solutions to similar problems. Think again about the tube-and-wing aircraft design, and how airport and aircraft design evolve in a symbiotic manner.

Designers might choose the easiest path to achieve a goal. This can translate to adopting widely accepted and proven design conventions to minimize resistance from users and manufacturing processes or lead to the adoption of designs that have been proven to work reliably and consistently. Many industries are fond of starting by adopting "reference designs" that save them the time and effort of starting from scratch.

Imitation is the ultimate flattery, they say, and it's very present in design. Mimetic isomorphism is the complicated term to describe the tendency of an organization to imitate another organization because the latter appears to be more successful or legitimate. In product design, this manifests as companies imitating successful designs from competitors to achieve similar success.

All this being said, and with the curse of unoriginality hanging on top of our heads, is there any room left for innovation, then, or what? There is, but such innovation cannot grow unaware of the curse. The only non-suicidal way is to be innovative while navigating through the proven architectures—which are proven for good reasons—as they have survived the test of time and countless iterations.

There are successful, high-profile cases of innovation on top of established architectural traits. SpaceX managed to introduce innovation to launch vehicles, even though rocket launcher architecture was established several decades ago, with the engine at the bottom, a set of staged sections that also host the tanks, and a payload at the tip. SpaceX introduced the concept of reusability in the first stage, which hadn't been done before. The innovation was done—not without failures—while keeping the overall architectural traits, somehow honoring and leveraging the decades of convergence of rocket engineering.

Engineers are constrained by the context to be unoriginal, to follow the established architectures that have formed with time, iterations, and are shaped by many factors they cannot control. However, this does not mean we are cursed to be boring and repetitive; our creativity needs to be focused not on creating something so radically new that it will never work because of sheer impracticality, but on creating stuff that inherits from something that exists and makes it a hundred times better.

John Gall said it beautifully decades ago:

All complex systems that work evolved from simpler systems that worked.

Chapter 23
The Strange Immortality of Failed Projects

Abstract Failed projects don't die; they agonize, mutate, and compete for resources with good projects. Bad engineering projects undergo many transformations except proper cancellation: rescoping, merging, rebranding, reassignment, or slow suffocation, but never clean termination. Aimless projects persist because killing them is tough. Sunk costs create incentives to continue, while the stigma of cancellation is (still) associated with failure and prevents decision-makers from wielding the axe.

Keywords Project management · Bad projects · Project cancellation · Rudderless projects · Life cycle stages · Sunk costs

In more than two decades working on all kinds of projects—some of which should have never ever seen the light of day under any reasonable circumstance—I have never had the pleasure to experience a project being properly canceled. I mean shelved, terminated, axed for good. I have seen projects being heavily rescoped. I have seen projects being merged, that is, opportunistically combining a group of ill-fated smaller adventures together to supposedly alleviate their despairing existence, only to multiply their misery. I have seen projects being rebranded, which is an internal stunt consisting of only changing their name in order to bring a breath of fresh air into the otherwise fetid atmosphere. I have seen projects being reassigned to another project manager with theoretical healing hands. And I have seen projects being left to rot and agonize, incrementally deprived of resources in the hope their flame will be extinguished, only to subsist against all odds. But I have truly never seen a project manager saying, "brothers and sisters, the dream is over, the project is *kaput*, let's focus on some other things".

Anyone who's spent a few years in the game has surely been part of an aimless project.

Statistically, an aimless project rarely gets back on track at any point in its life, no matter how long it runs or how many people are thrown at it. In the overwhelming majority of cases, these projects only get worse with time.

The feeling of being part of one is… peculiar. You sit through the meetings and try to paddle as hard as you can, but the atmosphere is thick with that quiet, shared

I. Chechile, *Engineering is Broken*, https://doi.org/10.1007/978-3-032-18169-5_23

knowledge that no matter how much effort we put in, the puppy isn't going anywhere. That it's just a matter of time until reality slaps us all in the face with an open hand. A slap we even want.

However, these projects don't die. Nope. Just when you think they're taking their last breath and you'll finally be free to do something meaningful with your life, they rise again like a deformed phoenix, flapping back into existence just to defecate a steamy one on your mental health.

Sailing through a project like this—especially if you're aware of its doomed nature—requires a special kind of character. It takes tremendous spiritual strength to get up on a Monday morning knowing that at 9 am you have a "progress" meeting, although there isn't any progress to talk about, just like in the previous 9 months. It requires keeping your internal brakes in good condition, because you can't just say what you really think; you nod along and let time do its thing.

An obvious problem pops up: how do you objectively diagnose that a project is terminally ill, besides the emotionally-charged answers you may get if you ask people around? The technocrat that lives in me wants to think it all boils down to metrics. However, it's not so simple. Are consistent budget overruns enough symptoms to kill a project? Not really, because it can be claimed that the project was initially underestimated by those who allocated the money in the first place. Skunkworks projects might be difficult to estimate if what's being pursued has little or no heuristics to compare with.

Consistent schedule overrun then? That must be it, right? Not really either: we all know everything takes longer than expected. And, in some cases, cool stuff might be late. See the James Webb Space Telescope, a marvel of engineering that took forever and cost many times more than estimated. By all cold metrics, this project should have been killed ages ago, but that would have deprived humanity of a fundamental tool for exploring the Universe.

It's always easy to find a way to justify why a project needs to continue, and that's why it's so hard to kill them. Building a narrative that justifies their existence is not hard.

And there is the famous problem of "efforts already made" or "pains already felt" that makes projects difficult to stop. A project that has consumed brain power, money, blood, sweat, tears, and time for a long period is more likely to be kept rolling to justify the ordeals suffered. And this is a positive feedback loop: every day a project of this kind is allowed to continue existing, more pain is felt, which makes it even harder to stop.

And last but not least, there is a question of pride: no project manager wants to add a canceled project to his or her track record. What transpires from this is how etched in our minds is the correlation between project cancellation and failure. It's easy to think a project's cancellation happened because the people in it were not up to the task. This is extremely shortsighted, as many conditions can dictate a project's fate that are completely outside of the hands of those embarked on it. Political changes, external funding changes, compliance issues, and many other things we can't control. In fact, in complex projects, we can control only a small number of variables.

We should perceive a project's cancellation as beneficial to the organization at large. Just like when we cut bad weeds in our garden to prevent them from competing with desirable plants for resources like water and nutrients, canceling a failed project increases the probability of success of other projects. The longer you let a project continue marching like a zombie, the longer you let it drain resources that could be going somewhere more useful, and the more you are not working on something that could be your winning ticket. The mysterious Archibald Putt reminds us that the likelihood of success of a project is still more a matter of luck than planning, that success or failure of a project should, therefore, not be used as the primary measure of a manager's competence. He also teaches us that when corporations are forced to choose between a person with demonstrated successes in small projects and one with a demonstrated failure in a large project, they more often than not opt for the person with a large failure. So, failure in projects can actually help you land your next job.

Death sentence is something barbaric that we humans are (still) learning to leave behind. There are clearly better means to impart human justice. However, when it comes to ill-fated projects—as inanimate inventions as they are, incapable of feeling any pain—there is nothing especially wrong with going somewhat medieval and keeping the guillotine sharp.

23.1 The Elves and the Project Manager

One of my favorite books as a kid was "The Elves and the Shoemaker", which is a classic fairy tale about a poor shoemaker who leaves his last piece of leather on the workbench overnight. Mysteriously, by the morning, a pair of beautifully crafted shoes appears. The story reveals that the work has been done by cute little elves willing to help the poor man. Customers love the shoes, and the shoemaker prospers as the elves return nightly to help.[1]

Every time I come across that book, a similar fairy tale develops in my mind, not about shoemaking but about engineering projects. I have been part of projects where evil elves visiting during the night doing shenanigans was the only plausible explanation for why certain things were happening. My version is called "The Elves and the Project Manager", and it goes like this.

Once upon a time, there was a project manager who had been given a straightforward task: deliver a new system in nine months. The requirements were clear and concise, the budget was adequate, the staff was competent, and everyone felt that level of motivation only brand-new projects can give. One evening, the project manager left his carefully prepared plans on his desk and went home, ready to start fresh in the morning.

[1] I still have the book, and I even brought it to Finland so I could read it to my kids. Reading it as an adult makes me think the shoemaker probably suffered from severe schizophrenia.

Big was his surprise when he arrived the next day; something strange had happened. The plans had been altered.

The Gantt chart had mysterious new nested milestones scattered throughout. The calendar had been flooded with new meetings, with recurring weeklies and a plethora of placeholders marked "Catch Up", "Huddle", and "Sync Up" that blocked entire afternoons. The requirements document was now covered in comments, asking for more model-based approaches and requiring executive sign-off for basically everything.

The project manager, somewhat puzzled but determined, began addressing the concerns. He attended the new meetings. He huddled and eventually caught up. Then, he went home for some well-deserved rest.

That night, the elves—balding, in their mid-40s, chain-smoking and wearing short-sleeved shirts with pockets full of pens—returned.

By morning, the elves had managed to tamper with the signature chain in documents, which now required approvals from six additional executives, always including someone on holiday; this was an European engineering organization, and the elves knew Europeans on holidays don't even remember they have laptops. New deliverables had appeared in the schedule, with titles such as: "Requirements Traceability Matrix", "Risk Register Update Report", "Mission Validation Specification Verification", and FMEAs required down to the last washer.

Night after night, the elves came back, unstitching everything they could unstitch. The team's inboxes would be flooded each morning with issue-tracking tickets from multiple email handles and with random topics, making it nearly impossible to filter them. Calendar invites continued multiplying; meetings to read emails aloud, tiger teams' formation review, and review boards that reviewed the reports of other review boards. The elves knew the golden rule of broken engineering: When in doubt, form a committee.

The elves were effective at bringing minutiae to high levels of attention. They would add comments to documents questioning font choices, margin widths, and whether bullet points should be circles or arrows. They would raise concerns about concerns, successfully distracting everyone from the actual work. Every closed action item would be reopened overnight.

One night, they left a note to the software engineers saying that they were far too talented to keep using a perfectly functional software tool they had been buying without problems from a third-party vendor. "You could build a better one, totally in-house, in no time", they wrote on the dev team's whiteboard. "It would only take a few months. It would be cheaper, more configurable. It's insulting you're buying this!". Needless to say, the engineers couldn't resist the teasing, so they severed ties with the supplier, burning all possible bridges in the process, and embarked on writing the tool from scratch in Rust.

But then, one morning, something changed. The project manager arrived at his desk to find something strange: nothing had been altered. The Gantt chart was exactly as he had left it. No new milestones had appeared overnight. No new tickets. The documents had no new comments questioning applicable documents.

The elves had left.

In the original fairy tale, the shoemaker and his wife stayed up one night to discover who their mysterious helpers were. Upon seeing the helpful elves working with their broken clothes, they were moved with gratitude and made them tiny clothes and shoes as gifts. The elves, delighted and feeling appreciated, danced away and never returned.

The project manager soon understood that the project had become self-sustaining in its misery. The team had adopted the elves' methods, and they no longer needed nocturnal intervention. People were now naturally adding extra clutter without being asked to. They were now scheduling pointless meetings by themselves. They kept arguing endlessly about trivialities because that was the new normal.

The elves had transformed this engineering organization into a stalling machine, dragging its corporate feet, taking weeks to do what used to take hours.

Delighted, the little bastards danced away and never returned.

Chapter 24
The Job Title Bingo

Abstract Job title inflation has transformed professional identity into a buzzword bingo. Organizations grant intricate titles as cheap substitutes for raises, while people concatenate nouns under their names like there's no tomorrow. The situation is a world-class mess where "senior" means absolutely nothing anymore, everyone is a manager, and reading job titles in business cards gives you a headache.

Keywords Title inflation · Job titles · Professional identity · Personal branding · Credential inflation · Workplace culture

There is an inflation of job titles in the workplace, and the engineering domain does not shy away from the silly trend.

Many organizations seem to have stopped caring and have relinquished any authority to standardize and regulate the sentences they give to their people to put under their names on business cards and social media. Others grant titles as a replacement for monetary compensation.

And we, mortal, insecure featherless bipeds as we are, seized the opportunity to decorate our egos with lengthy, pointless piles of nouns and adjectives next to our names, to the point where nothing makes sense anymore.

The situation is bizarre. At the time of writing, everyone is a manager. A senior manager, even if it's their first job. I once saw someone in a company I was working for go from "intern" to "specialist" in one single profile edit. Meteoric career progress right there, which of course did not match their work reality because they were still doing intern stuff. Time ago, I had a call with a recruiter who was searching for a "Senior Product Manager". Five minutes into the call, she admitted they were looking for a backlog maintainer. That's something a Junior Product Manager can do. The senior word means absolutely nothing these days.

But hey, feel free to try the title bingo. The game is about creatively combining terms such as "product", "project", "delivery", "customer", "sales", "analysis", "operations", "business", and "solutions" in a salad of words that can give plausible things like "operations sales manager", or "sales operations manager" depending on

I. Chechile, *Engineering is Broken*, https://doi.org/10.1007/978-3-032-18169-5_24

someone's taste. "Head of" and "Global Head of" are also in order if you want to give it a halo of extra importance.

But it can get worse. Some people shamelessly put things like "visionary" and "disruptor" next to their names. There are people out there with titles like Chief Happiness Officer. It's pure madness. I have also seen titles with weird, broad scopes, like "Head of Technology and Business Development". I mean, I am all for polymathy, but that's a wide spectrum. Chances are they suck at both.

And the ones with a few sandwiches short of a picnic are the ones with the infamous vertical bar to concatenate titles and other terms like there is no tomorrow: Chief Architecture Manager | Guru | Advisor | Innovator. Some may include their previous gigs in their title, like | ex-ACME | former Whatever, as if anyone cared where they worked before. Some others add random, abstract nouns like "Leadership | Excellence | Culture". And there's also the ones with emojis in their title, like little rockets, fists, hearts. We will realize at some point in the distant future the damage that the little rocket emoji has caused to society.

Another popular stunt these days is half-quitting a job and become an "Advisor", "Special Advisor", "Strategic Advisor", or similar, highly preferred by semi-retired, gray-haired professionals, who manage to stay on the payroll part-time while they're employed elsewhere. Ka-ching. The more I browse LinkedIn, the more I realize I am the only idiot who actually quits jobs.

The engineering profession should know better. Like, come on. We have Staff Engineers and Senior Staff Engineers and Principal Engineers and Distinguished Engineers, and Fellows; I have even seen the words "Emeritus" used in some people's roles, which were corporate roles, not academic. It's all so tacky.

What motivates this madness is an unhealthy combination of institutional laziness and individual insecurity. Companies lack the nerve to tell people no, to maintain standards, to explain that you don't become a senior anything just because you want to or because you have been around for 6 months. Saying no requires determination, but it surely invites conflict, and modern corporate culture has been shying away from conflict at all costs, even when the cost is the complete erosion of meaning in the language we use to describe ourselves and our work. So instead of no, we get yes; we get title inflation as a substitute for difficult conversations, we get promotions in name because they're cheaper than raises and easier than feedback. Easier than earning it with hard work, mistakes, and time.

We have grown in a culture of hollow personal branding and professional anxiety where our worth seems to be determined by how we are perceived, and that perception is shaped by titles and credentials, and that falling behind in the title rat race feels like professional failure. So, we grab for every title upgrade we can get, we massage our job descriptions to sound more impressive, we add the vertical bar, we concatenate nonsense because everyone else is doing it.

The result is this semantic shitshow where words mean nothing because they mean everything, where titles have grown so long that by the time you stop reading, you already forgot the beginning. They no longer communicate anything useful about what someone actually does or how competent they are at doing it. When everyone is a senior principal distinguished fellow advisor emeritus, then no one is, and we're left trying to decode actual capability through data mining, because the titles themselves have become noise.

Liberace, the famous American pianist, singer, and actor known for resembling the exact opposite of anything you could reasonably call minimalism, a man who performed in a mirror-tiled Rolls Royce while wearing a costume that cost more than a house, once observed: "Nakedness makes us democratic; adornment makes us individuals". He understood something deep about human nature and our relationship to display, about how we use decoration to distinguish ourselves from the undifferentiated mass of other humans, to boost status and identity and worth. But Liberace was also a supremely talented entertainer whose adornment, though excessive, was in service of his craft, not a substitute for it. Our professional adornments aren't enhancing anything because there's often nothing underneath to enhance. We're wearing the costume without learning any instrument, hoping no one notices that beneath all the glitter, we're just another mediocre practitioner desperately hoping to be taken for something more.

At any rate, every organization is exclusively staffed by managers. This is a paradoxical organizational state: the moment everyone is a manager, no one is. An organizational absolute zero, where no work is done because there is no one left to lay a single brick.

In some ways, managerial inflation is understandable: as growth ramps up, more people come in, more "processes" appear, more bureaucracy, more to be managed: people, papers, machines, systems, projects, products. The crowning spree goes brrr because there's an increasing urge to encapsulate and package complexity behind single faces.

Crowning managers is like printing money: easy to do, but the implications can be uglier than the problem it may come to solve. As managers are minted much faster than leadership can be grown or taught, the result is people with lengthy titles but low legitimacy, who are incapable of making any relevant decision, being left to float in a sea of peripherality and administrative chores.

The path out of this mess is kind of utopic, because it requires organizational discipline and individual honesty, both of which are harder to find than the graviton. For starters, organizations need to grow a pair, establish real standards for what titles mean, and defend those standards even when it makes people uncomfortable or causes them to leave for places with looser rules and easier ego massaging. Individuals need to get comfortable with career accuracy, with describing what they actually do rather than what they wish they did, with the notion that their worth as humans and professionals isn't determined by the number of nouns and adjectives they can chain next to their name. We need to remember that engineering, at its core, is about building things that work and producing value, neither of which requires a LinkedIn profile that reads like a tongue-twister. The best engineers I have ever worked with had very simple titles and extensive accomplishments. The worst had glamorous, convoluted titles and no technical depth whatsoever. That's not a coincidence. Until we collectively decide to stop lying to ourselves and each other about what we do and who we are, we're going to keep drowning in a sea of meaningless words, stupidly decorating ourselves with empty titles while the actual work goes undone and the problems that need solving remain unsolved.

Chapter 25
The Kindergartization of Work

Abstract At some point, someone decided that making professional adults do the same things kindergarten kids do could fix dysfunctional teams. Managerial infantilization has filtered into the modern workplace, where instead of confronting the problems that make work difficult, organizations transfer their handling to facilitators armed with silly practices, markers, laser tags, and camping tents.

Keywords Team building · Workplace culture · Human resources · Team bonding · Jelled teams

Required materials: several large sheets of paper and permanent markers. Give each participant a marker, clear out enough space for people to walk around (remove chairs and tables), and, optionally, put on some background music to set the mood.

This is the Poncho Game, designed to facilitate team building. Also known as "The Gratitude Poncho", the game consists of writing positive feedback to our colleagues on their paper ponchos. But let's get to what really matters: how to make a gratitude poncho.

First, fold the paper sheet in half, then cut the corners into a rounded shape (be generous with the hole so colleagues with more voluminous heads can also fit), and voilà: your poncho is ready to receive the best feedback imaginable. Put it on and start the game.

Then, just follow your instincts and let the game "flow" as long as it feels right. After enough time, pair people up, take off the ponchos, and reflect on the feedback written on them. When the poncho game ends, the team will feel more united and resilient.

Record Scratch

Wait, what?

Okay, banal practices have always existed. And let's be fair: there's no need to blame the psychologists or coaches who have figured out how to make a living catching lazy managers off guard and selling them "team building" workshops of the most ridiculous kind. Everyone's got to put bread on the table.

I. Chechile, *Engineering is Broken*, https://doi.org/10.1007/978-3-032-18169-5_25

The real blame goes to those who actually believe—or pretend to believe—that an exercise like the "gratitude poncho" will turn a mediocre team into a good one. That it will turn a broken, demotivated team into a cohesive and efficient unit.

The infantilization of workplace relationships we seem to be witnessing has a somewhat understandable purpose: to avoid the complicated work of making a team function well. The logic is like this: as a boss, I do nothing, but if finance approves, I'll hire a couple of "facilitators" who'll bring in a "fun" exercise, and bam, the team will start working better.

Also, what is with this obsession with forcing a team to "have fun"? What is it with this managerial fixation on dictating when a team should make merry? Aren't work teams made up of adults who can decide for themselves when they want to do something enjoyable together?

If a team hasn't had a single after-work hangout in years, fine, maybe that might signal something.[1] But if they don't do anything for a couple of weeks or months, is that really a sign they need to be thrown into an escape room or laser tag? We seem to be witnessing a kind of *kindergartenization* of the workplace culture, as if employees were kids at the end of elementary school writing on each other's uniforms.

In a past life, a boss tried to send me on a sort of "spiritual retreat" with other managers so we could "connect" better. This retreat involved camping. I am not a camping person. I remember asking myself: Why on earth do I have to share a tent with the COO to make us work better together? Isn't it enough to hash things out face to face, Monday to Friday, 9 to 5, as many times as needed if something isn't working? While not exactly a fan of meetings, it's a better way to sit down in a meeting room to smooth things out with this colleague than in the middle of nowhere, with no running water and mosquitoes flying all around. I don't want to listen to Creed from Quality Assurance snore or fart at 3 a.m.

Managers will send people go-karting or make them wear a gratitude poncho, but they won't address the real, underlying issues that actually break a team: favoritism, nepotism, abuse, and harassment. That, they won't fix.

[1] Recently, I was visiting a governmental facility in Espoo, Finland. Someone was giving a tour of the facilities to the group I was part of. At the end of the tour, there was a short Q&A, and someone from my group asked a (somewhat odd) question: "Do you have parties with your colleagues?", to which the man answered: "We haven't had a party in like 25 years".

Chapter 26
Let It Rot

Abstract Everything that comes into existence needs our sustained attention. As engineers, we shouldn't forget about the importance of following things up, getting them done, closing them. Engineering, in general, is a lot about planning and determinism, but also about expecting the unexpected; things pop into existence out of the blue and often at the worst time possible. Whatever their nature, all things have a life cycle, from appearance to closure. And things do not seem to close by themselves.

Keywords System life cycle · Task follow-up · Task management · Task closure

There's something so basic yet so underestimated in engineering about follow-through, understood as the ability not just to start something, but to monitor it until it's finished, or at least reaches a meaningful level of maturity.

We're experts at throwing new things into the air. Ideas, suggestions, changes, problems, modifications, initiatives about this or that. But only few people actually take responsibility for the entire life cycle of what they have started. For making sure that what began reaches some kind of end. For coming back to the seemingly trivial thing they threw in a meeting that you thought they would instantly forget.

Engineering departments—and perhaps tech startups being the most emblematic case—are places of serial initiation, not of follow-through, and even less of actual completion. You have most probably seen a big new thing show up out of nowhere on a Monday morning, announced with fanfare, and accompanied by the obligatory Gantt chart someone rushed over the weekend, only to let it rot forever after.

We have witnessed someone, with solemnity, spin up a brand-new issue-tracking board and then abandon it to die of neglect.

One of my former bosses, famous for his impulses and pulling ideas out of thin air, once assigned me something fairly complex. Nervous, I asked a colleague for advice. This colleague said:

—Don't worry. He'll forget.

And he forgot. He was notorious for never following through on what he started.

I. Chechile, *Engineering is Broken*, https://doi.org/10.1007/978-3-032-18169-5_26

I'll admit a guilty pleasure: browsing through forgotten documents, spreadsheets, Gantts, and tickets as if they were relics of a past that's never coming back, remembering with a touch of nostalgia the aura of ephemeral importance that accompanied their birth. Once declared the most critical thing that ever existed, they now gather dust in folders whose last visit is from years ago. There's something strangely captivating about the contrast between that glorious beginning and their inevitable fate of abandonment.

And it makes sense. Starting something is easy. Throwing a new idea on the table takes little energy. But following it through—making sure it reaches a destination—requires time and dedication, and no one seems to have either.

I think: if we let documents, Gantts, or issue-tracking tickets be forgotten—things that at least exist and have some tangible entity and visibility—what chance do those other things have, the ones that don't even get written down? Things tossed into the air during a meeting or a hallway chat. Their only destiny is to be buried by whatever comes next, only to be forgotten too.

Closing things, getting them to completion, undoubtedly requires you to give a damn.

In the engineering practice, it's becoming increasingly rare to find people who truly give a damn. In reality, what you more easily find are the zero-damns-given kind, unbothered to sustain in time any mental engagement. "Not my problem" kind of souls.

In the 1920s, Bluma Zeigarnik noticed that waiters in Vienna could flawlessly recall unpaid orders but immediately forgot anything that had already been settled. Unfinished tasks, she discovered, create a sort of mental itch; the brain keeps them alive until they reach closure. It's called the Zeigarnik effect.

Strangely, the engineering world seems to have developed an immunity to it. Our unfinished tasks don't haunt us. The itch never comes. We seem to be the only species on Earth that can start something, forget it after a while, and then be surprised six months later when someone asks, if anyone ever asks.

Chapter 27
The Recurring Cast

Abstract Every engineering organization is an evolving social experiment, where a diverse blend of personalities emerges with a certain predictability. These are archetypes that arise from the conditions that modern engineering organizations create. You might recognize every character described here, either because you have suffered them or maybe because you're one of them. Understanding why so many engineering organizations struggle to produce actual work requires unpacking their social scene, their balance between the population of these personalities, and how these personalities have become essential in making engineering every day a little bit easier or more difficult.

Keywords Workplace personalities · Assholes · Cronies · Wannabes · Rainmakers · Technocrats · Floaters · Anti-skilled

Every engineering organization of a certain size eventually develops its own cast of recurring characters. Personalities that appear with certain predictability. We find the full spectrum: annoying ones like floaters, assholes, wannabes, rainmakers, and hierarchy acrobats. But also, pleasant and essential ones like tightrope walkers or technocrats. You might recognize every character described here, either because you have seen them in the wild or maybe because you're one of them. You might find people with combinations of these traits. Note that there are probably more than these, as the social continuum is not so easily discretized, so I will leave it as an exercise to the reader to recognize other personalities to add to the list.

27.1 The Floaters

The art of doing the bare minimum while avoiding making it look like underperformance is one of the most perfected disciplines in the global corporate ecosystem, and that includes engineering departments. Not everyone can sit in the same chair for a

I. Chechile, *Engineering is Broken*, https://doi.org/10.1007/978-3-032-18169-5_27

decade without anyone asking what they actually do. It takes skill. Important clarification: doing the minimum isn't precisely laziness. It's the skill of minimizing energy expenditure, or maximizing the time for completion, and in the process blending with the background.

The floater isn't really burned out or stressed. Why wouldn't they be? Floating embodies the calm wisdom of someone who came to the conclusion a long time ago that working hard is a move with little return on investment. While others scramble around, desperately trying to stand out, the floaters coast peacefully. While objectives and deadlines are written, and nights are wasted on PowerPoints no one will read, the professional floater glides with the ease of someone who has already played the game and decided it's not worth running for a promotion that only brings more emails and more meetings. They understood that the more a boss gives you, the more it entitles them to ask you in return.

Every office has at least one. Socially skilled, they specialize in small talk and in positioning themselves right at the center of where nothing happens. The kitchen is their spot: that's where they talk politics, football, and the economy, stretching coffee breaks into infinity. Remote work during the pandemic nearly wiped them out, but they returned stronger than ever.

The system rewards the floaters with stability because they don't do harm. They question nothing. They bring no drama. And that's quite a good list of traits in these trying times with high-maintenance special snowflakes being born every minute.

Many of them could do more. Some of them were and still are technically brilliant. Some are not so bright. However, they all understood that energy is a non-renewable resource in a world that loves to waste it in posing and careers built on friendships and artificial visibility.

What's interesting is that many work environments prove them right. Workplaces where error is punished more harshly than inaction produce coasters in mass by disseminating the idea that indolence and feet-dragging is the best protection against the surrounding bullshit.

27.2 The Assholes

Have you ever thought "Wow, what an absolute asshole!" about someone? I bet you have. I have. But who deserves to be branded as an asshole? Many of us use the term indiscriminately, applying it to anyone who annoys us, gets in our way, or happens to be enjoying greater success than us at the moment. But a precise definition is useful if you want to spot them in the wild and avoid them. It can help you distinguish between colleagues (or customers) you simply don't like from those who actually deserve the label. It can help you distinguish people who are having a bad day or a bad moment ("temporary assholes", as it isn't fair to call someone an asshole based on a single episode) from persistently nasty and destructive jerks ("certified assholes").

Researchers who write about psychological abuse in the workplace define assholery as "the sustained display of hostile verbal and nonverbal behavior". That definition is useful as far as it goes, but it isn't detailed enough for understanding what assholes do and their effects on others. Assholes hit a special, unpleasant nerve as they extract every ounce of joy from our system.

Bob Sutton, in his evergreen "The No Asshole Rule", suggests a simple test to spot them:

- Test One: After talking to the alleged asshole, do you feel oppressed, humiliated, de-energized, or belittled by the person? In particular, do you feel worse about yourself than before talking?
- Test Two: Does the alleged asshole aim their venom at people who are less powerful rather than at those people who are more powerful or equals?

There are many other actions—sociologists call them interaction moves or simply moves—that assholes use to demean and deflate their victims. Dr. Sutton goes further and lists twelve common moves ("the dirty dozen", as he calls it) to illustrate the range of these subtle and not-subtle behaviors used by assholes:

The Dirty Dozen

1. Common Everyday Actions That Assholes Use
2. Personal insults
3. Invading one's "personal territory"
4. Uninvited physical contact
5. Threats and intimidation, both verbal and nonverbal
6. "Sarcastic jokes" and "teasing" used as insult delivery systems
7. Withering e-mail or Slack flames
8. Status slaps intended to humiliate their victims
9. Public shaming or "status degradation" rituals
10. Rude interruptions
11. Two-faced attacks
12. Dirty looks
13. Treating people as if they are invisible

I Would Add a Few More:

1. Shamelessly throwing people under the bus when things go bad and equally shamelessly taking the credit when things go well
2. Tone deafness coupled with a consistent incapacity to "read the room"
3. Lack of awareness or interest in building rapport
4. General incapacity to be on top of things (out of laziness, incompetence, or the pretense that they're too busy or too important to be on top of things). This creates a sense of constant dissatisfaction that, in their asshole minds, justifies them to be jerks.

It takes some effort to qualify as a certified asshole: a person needs to display a persistent pattern, to have a history of episodes that end with one target/victim after another feeling belittled, put down, humiliated, disrespected, oppressed, de-energized, and generally worse about themselves. Psychologists make the distinction between states (fleeting feelings, thoughts, and actions) and traits (enduring personality characteristics) by looking for consistency across places and times if someone consistently takes actions that leave a trail of victims in their wake, they deserve to be branded as certified assholes.

We all have the potential to act like assholes under the right (or should we say wrong!) conditions when we are placed under pressure or, especially, when our workplace dynamics encourage those in power to act that way.

Although it is best to use the term with care, some people do deserve to be certified as assholes because they are consistently nasty across places, times, and situations.

It is also important to define the term asshole because this is not an argument for breeding spineless wimps; there is virtue in healthy conflict, even noisy arguments. Research on everything from student groups to top management teams reveals that constructive arguments over ideas—but not nasty personal arguments—drive greater performance.

The right kind of friction can help any organization. There is good evidence that spineless and obsequious wimps damage organizations. Experiments and field studies in organizations show that when teams engage in conflict over ideas in an atmosphere of mutual respect, they develop better ideas and perform better.

And let's also put a good word in for socially awkward people, some of whom—through no fault of their own—are so socially insensitive that they accidentally come across as assholes at times. Certainly, people with high emotional intelligence who are skilled at taking the perspectives of people they encounter and responding to their needs and feelings are pleasant to be around and well-suited for leadership positions. However, many extremely valuable employees—as a result of everything from being raised in dysfunctional families to having disabilities like Asperger's syndrome and nonverbal learning disorders—act strangely, have poor social skills, and inadvertently hurt other people's feelings.

27.2.1 The Making of an Asshole

So far, we have been talking about certified assholes who are already asshole-ing full-time in a steady state. What is their journey to get there? Why do assholes become assholes? Is it a gradual thing, or are they assholes right out of the womb? Or is it from kindergarten age? Believe me, I have met a few little proto-assholes in my kid's kindergarten. There was this one kid who wouldn't stop kicking, biting, pulling other kids' hair, and consistently disregarding personal space, even in front of his parents, who would not move a finger to curb that behavior. Although his parents had not transpired as assholes to me, their inaction was, in my view, sowing the seeds of a life of assholery by not letting their (single) child see that his actions were unacceptable. I have no doubts that this kid, should he get into any power position in his adult life, will become a certified asshole 25 years down the road.

And it's not just me again doing my typical dilettante psychological analysis here. Behavioral psychology suggests that people learn behaviors through observation and reinforcement. If someone grows up in an environment where dominance, aggression, or belittling others is rewarded or at least not punished, they may internalize these behaviors as effective strategies for achieving their goals. Social learning theory highlights how individuals imitate behaviors they see rewarded in others, particularly if those behaviors lead to status or success.

Personality traits play a significant role, particularly those linked to dominance, low empathy, and competitiveness. Research in personality psychology suggests that traits such as narcissism, Machiavellianism, and psychopathy—together known as the Dark Triad—correlate with exploitative and nasty behaviors. While not everyone with these traits becomes toxic, individuals who exhibit high levels of these characteristics are more likely to engage in manipulative or demeaning behaviors, especially if their environment tolerates or rewards such conduct.

Of course, cognitive biases and self-justification mechanisms further entrench these behaviors. People tend to rationalize their actions to protect their self-image. A person who repeatedly mistreats others may convince themselves that their behavior is justified, believing that the world is harsh and only the strong survive. This mindset can be reinforced in workplaces where success is often linked to assertiveness, and some individuals interpret this as permission to act aggressively and unemphatically.

The role of power is another significant factor. Give power to an asshole, and you have a dangerous individual who can inflict damage on their subordinates. People in positions of authority often receive less critical feedback and may become insulated from the consequences of their actions. This diminished situational awareness makes it easier for them to act in ways that harm others without feeling guilt or remorse. Often, assholes can be carefully selective, conveniently acting like jerks to those who can't defend themselves and act nicely—and even show submissive behavior—with those who are either too important for them to succeed or simply more powerful. It's always interesting to see an interaction between a full-time asshole and a Single Point of Failure (see a few sections below); assholes are unusually nice to them.

Workplace and cultural environments also determine whether someone fully embraces and maintains toxic behavior. If an organization tolerates or even rewards aggressive, demeaning leadership styles—either explicitly or through unspoken norms—individuals are more likely to continue these behaviors. Toxic leaders can create cycles where employees, in turn, adopt similar behaviors to fit into the culture or to protect themselves.

Ultimately, a "certified asshole" is not simply born but shaped through a mix of personality predispositions, social modeling, reinforcement, and the structural dynamics of power and culture. While some may have a stronger inclination toward such behaviors due to innate traits, their persistence and escalation depend on whether these behaviors are met with resistance or rewarded.

27.2.2 Can It Get Any Worse? Yes, Assholes Might Team Up

When two or more "certified assholes" with power align, it's like a bizarre orchestra of assholery playing a long, awful song out of tune. A clique of certified assholes validates each other's behaviors, reducing any self-doubt or external criticism they might have otherwise faced alone, making them more resistant to feedback and correction.

They may openly manipulate organizational structures, silence dissent, and enforce compliance through fear or coercion. This is why toxic leadership is difficult to uproot; when one powerful asshole is supported by another, they create a protective network that repels accountability.

When a single toxic leader enacts harmful policies or mistreats subordinates, there's a clear source of blame. But when multiple toxic individuals reinforce each other's actions, responsibility becomes a bit diluted. Each member of the toxic alliance can point to the others as co-decision-makers, making it difficult to pin wrongdoing on any one person.

This dynamic is further strengthened by in-group favoritism and out-group hostility. People in positions of power too often form exclusive, even bizarre in-groups with colorful characters, perceiving those outside their clique as inferior or even threatening. When two or more assholes with power associate, they create an elite inner circle where loyalty to the group matters more than ethical considerations.

The combination of mutual reinforcement, power insulation, responsibility diffusion, and in-group bias turn a toxic leader into a toxic system.

27.2.3 Leadership by Erasure

Of course, certified assholery does not go unnoticed for too long, and, sooner or later, people react to it. It might be low-key or open, but consistent asshole action eventually creates a reaction. Whether a certified asshole knows they are an asshole or not, it depends, but the environment will eventually let them know. And the key here is what will the certified asshole do with this information.

In rare cases, they might correct or tone down their behavior. However, being so tied to personality traits, most assholes remain assholes for their lifetime.

In cases, assholes in power may choose what we can call leadership by erasure, which means erasing/removing/firing/sidelining/letting go of whoever has realized their assholery and replacing them with people who haven't had (still) the time nor the pleasure to detect it, or are just functional idiots who just yes-boss their way in life. Of course, because a proper certified asshole is a perpetual alienation machine that even violates thermodynamics laws—it alienates without any input of energy—they will sooner than later alienate the newcomers as well, so the method requires a constant supply of fresh new souls; a revolving door to dispose of the damaged ones and replace them with fresh, blissfully ignorant people who have no clue what they got themselves into while they believe they have just landed themselves their dream job.

27.2.4 Post-Asshole Life

Life after escaping the gravitational pull of a certified asshole (or worse, a whole team of them) can be a strange, almost surreal experience. At first, there's the detox phase; a mix of relief, confusion, and a weird silence where the constant stress used to live. If you have spent months or years in a toxic environment, you might catch yourself bracing for nonexistent criticism, overexplaining simple decisions, or hesitating before speaking; symptoms of what can only be described as "asshole-induced PTSD". It's as if your nervous system is still waiting for the next passive-aggressive Slack message, the next condescending smirk.

Recovery involves learning how to function again in a world where kindness isn't a trap and respect doesn't come with a hidden power play. The first step is unlearning the survival mechanisms you adopted under the rule of an asshole. You don't actually need to second-guess every interaction. You don't have to preemptively take blame for things you didn't do. And no, not everyone around you is secretly plotting your downfall; some people just exist without turning life into an endless power struggle. Wild.

Then comes the rebuilding stage. This is where you reclaim lost confidence, reestablish healthy boundaries, and remember what it's like to work or live without constantly navigating landmines. There's a certain joy in realizing that, yes, it's actually possible to work, create, or lead without belittling others or walking on eggshells. That human interaction doesn't have to be a series of childish dominance games.

Of course, there's also the lingering bitterness to deal with. The reality is that some assholes never face consequences, and nothing short of a cosmic reckoning will fix that. However, the best revenge—as cliché as it sounds—is living well.

27.3 The Wannabes

Just because you badly want to be in an industry does not mean you have anything to offer that industry. I would love to work in an F1 racing team. But could I? I have got nothing to offer except my appreciation for the heavily logoed gear, the fancy headsets, and the big watches everybody seems to be wearing.

Not everything I consider cool is worth having me in it.

There seems to be an accelerating proliferation of people with childlike dreams and the unstoppable urge to be in certain technical industries but lacking the most basic skills or competence to offer anything to them. Fueled by hype, they bring enthusiasm and fascination but no expertise. Interestingly, some of them find their way up the ladder and influence the lives of those who do have expertise and qualification, making a wannabe also a close sibling with the hierarchy acrobat (more about that soon).

The tech industry is particularly welcoming with specimens who fell in love with the sector's prestige and its sci-fi aesthetics without ever bothering to understand what they're actually dealing with. They read a few articles on the topic, watched some videos on YouTube, and thought they were ready to have a leadership position in an engineering organization by exploiting their social engineering skills, because that's all they have to offer.

These creatures are particularly damaging when they have actual power. They make decisions that affect the people doing real work. They mandate changes based on the cloud of buzzwords that rule their thoughts at that moment—they think in buzzwords—randomly collected at conferences whose cocktail parties they never forget to document in social media. They define roadmaps and strategies for technologies they don't understand. They restructure technical teams, destroy everything that took time to put together, and alienate everybody in the process.

You can spot them by their vapid sci-fi vocabulary. They tend to speak about "revolutions" and "eras" of X or Y. They jump from wagon to wagon, from fad to fad as they get bored, like my kids switch from Ninja Turtles to Transformers.

See, there's nothing inherently wrong with being aspirational. Nothing too wrong either with liking an industry and wanting to be part of it. We were all rookies at some point. Everybody's got to start somewhere. However, there's a substantial difference between starting at the bottom to learn and painstakingly build a career with effort, versus cutting the line entirely because you liked the view and you happen to know the right people. The rookie knows they don't know and works to learn. The wannabe deceives everybody but ultimately lies to themselves. A lie that seems to protect them from feeling any shame when talking about what they do not know.

27.4 The Rainmakers

Some chapters ago, we talked about bluffing and the art of selling vaporware. Well, this personality type is the epitome of all that. If you ever happen to work close to one of these specimens, watch them carefully, for they are a marvelous creation of the corporate ecosystem.

Born in 1891 in Concepción del Uruguay, Argentina, Juan Baigorri Velar's early years were unremarkable. He trained as a geophysicist, studying in Europe and gaining an education in atmospheric sciences.[1] But his return to Argentina in the 1930s marked a shift from conventional science to something closer to showmanship. Baigorri began touting an invention that, by his own account, could bring rain anywhere, anytime. This device, he claimed, manipulated "electromagnetic fields" (sic) to influence atmospheric moisture and induce precipitation.

Of course, the scientific community demanded proof. Baigorri, an elusive figure, refused to disclose the inner workings of his gizmo, citing fears of intellectual theft. What he did offer, however, were public demonstrations that would boost his legend. He only became more and more famous.

Baigorri's so-called rainmaking machine was described as a modest, box-like contraption, the kind of device that might not look out of place at a magician's show. It was small, unassuming, and wrapped in a fog of technical mystery. Baigorri's confidence in its efficacy, however, was anything but small. He traveled to drought-affected regions, promising rain to desperate farmers, and, from time to time, he appeared to deliver.

One of his most famous stunts took place in Santiago del Estero, a province devastated by drought. With crops failing and livestock dying, the region's people had little to lose by listening to the claims of our friend. Baigorri declared that he could bring rain within days. Miraculously—or coincidentally—heavy rainfall followed. Crowds hailed him as a savior, while skeptics dismissed him as a salesman who had simply gotten lucky.

Baigorri's rise to fame followed a familiar pattern: make bold, unverifiable claims and exploit some pressing need. Claims need proof. It's the same playbook used by countless sellers of "revolutionary" solutions throughout history. From Charles Hatfield, the American "rainmaker" who allegedly used secret chemicals to coax rain from the sky in the early twentieth century, to modern corporations that promise disruptive technologies only to deliver nothing, but rake in millions of investors' money in the process. One thing's for sure: there isn't a drought of gullible investors. Elizabeth Holmes managed to swindle Henry Kissinger, a seasoned war criminal who worked closely for decades with the CIA overthrowing governments around the world. You wouldn't think Mr. Kissinger was precisely easy to fool. And yet she did lure him into the Theranos nonsense, ripping millions off him.

What makes figures like Baigorri so attractive (and frequent) is their ability to tap into sensitive human needs. Argentina in the 1930s and 1940s was no stranger to drought and agricultural crises, and the rural poor were eager for a solution, no matter how improbable. For those communities, Baigorri was a symbol of hope.

Hope and fear are easily exploitable commodities. Fear also sells; at the time of writing, rainmakers are tapping into defense and military applications due to "growing geopolitical tensions" they love to stir up while seeking to ride on rising defense budgets. Blame the game, eh?

Baigorri's rainmaking device, much like modern vaporware, was always just convincing enough to keep people believing. He offered no verifiable data, no detailed

[1] Several details about Baigorri's background are doubted as false, so it's not entirely clear where he studied and what exactly.

explanations, and no repeatable results. Instead, he relied on the psychological power of belief: if it rained after he flipped a switch, then surely he had made it happen.

To his credit, Baigorri never shied away from the spotlight. He accepted challenges from skeptics, including a particularly memorable one in Buenos Aires, where he promised rain in the Villa Luro neighborhood. Rain did fall, and Baigorri walked away triumphant.

The scientific community remained skeptical. Meteorologists pointed out that his successes could be attributed to chance or seasonal patterns. Even the timing of his demonstrations, critics noted, often coincided with forecasted precipitation. Baigorri doubled down, dismissing detractors as "envious" or "ignorant". His refusal to allow independent testing of his device only added to the skepticism surrounding him.

As the years went on, Baigorri's star began to fade. His refusal to provide evidence alienated even his supporters. The world moved on, leaving Baigorri to languish in obscurity. He died in 1972, taking the secrets of his rainmaking device to the grave.

The rise and fall of our friend Baigorri seems to mirror the trajectory of countless other "disruptors". Today's world is full of similar slippery figures promising revolutionary products that don't exist and will never exist. Self-proclaimed "visionaries" claiming unverified solutions to urgent global problems that will keep appearing as soon as there are fools ready to throw money at them, come rain or shine.

27.5 The Crony

Throughout its evolution, every organization eventually gives rise to a very particular character, one that finally deserves to be brought to the spotlight: the crony. Also known as the buddy, or the org chart acrobat.

The crony is that person who somehow manages to be part of every project that emerges, always with respectable-sounding titles empty of actual substance or relevance. A world champion in strategic friendship and yes-ism as their main career development tools, the crony is selectively close to those who make decisions, which ensures they always land on their feet, project after project. In cases, although not always, the crony is a high-school or childhood friend of one of the bosses, a condition they milk to the last drop.

The crony positions themselves as someone trustworthy, offering loyalty to their sponsors, though they're not necessarily explicit bootlickers. They know how to navigate that thin line.

Important clarification: the crony should not be confused with a climber. The climber seeks status, power, and influence, and for that they need to climb higher and higher in the hierarchy, no matter what. The crony, on the other hand, is intellectually incapable of such Machiavellian synapsis. They seek comfort, visibility (this is important for ensuring the next move), and survival. The crony will occupy their spot for as long as the project lasts, be it 2, 5, 10 years or more.

The crony brings no real technical value. But that's not a problem, because they don't seek direct responsibility or skin in the game. Others will do the heavy lifting, not them.

The crony's mission is to be *there*. They weave their web of connections, knowing well that this network is the key to securing a spot in the next big thing. They know how to insert themselves into org charts like a default setting, appearing as the natural choice to fill empty "ravioli" boxes that need a name. Sometimes it feels they have hacked the template file of the org chart, and that every time someone starts a new one, their face is automatically there with a proper title out of a buzzword bingo.

Like an extra in an action movie, the crony fills the frame but adds little to the plot.

The crony is friendly and has good "soft skills". They don't make trouble or create drama, because drama isn't good marketing for future projects. Charming, chatty, affable, always with a joke ready to break the ice, the crony calls for meaningless meetings whose only purpose is to remind everyone they're part of the cast.

They're a permanent fixture in "important" meetings, having also hacked mail clients' autocomplete feature with their name always being suggested in every CC field for reasons no one can comprehend.

The crony travels. Conferences, trade shows, client visits, supplier meetings. That's what sets them apart from floaters. The floater seeks routine and calmness; the crony is fine moving around the world, accumulating mileage. Their LinkedIn profile photo is always either Dubai, Paris, or New York. You can always spot them in the company's official photos, hands neatly clasped in front, smiling, always wearing their lanyards.

The crony volunteers to staff the company booth at every expo they can, cheerfully chatting with random passersby, handing out brochures, networking. The crony frequently makes to the company's social media channels. Collaborative, the crony happily participates in corporate videos and testimonials.

The crony's professional past is as zigzagging as their role is hard to define. Like a footballer changing clubs as often as underwear, the crony adapts. They seek no consistency in their career path, leaving behind a trail of projects of every shape and color. Professional heterogeneity is their jam, and one of the easiest ways to spot them. If you ever want to find the crony during a review with customers, go check the place where the sandwiches are, and you will find them nodding with a smile and ready to kick off a friendly chat.

The crony thrives thanks to the fact that many organizations have stopped considering actual contribution as a figure of merit. In the absence of any metrics for engineering value, hierarchical dexterity becomes an important skill.

27.6 The Tightrope Walkers

Let's stop for a moment the saga of weird personalities that add little to no value and talk about something a bit nicer. So, let's talk about the tightrope walkers, a *rara avis* in engineering domains. A close cousin of the technocrat we will discuss in a section below, the tightrope walker doesn't necessarily need to be technically brilliant but be a generalist who can tie the loose knots.

The tightrope walkers smooth out the rough edges, whether it's a cryptic message handed down by a higher-up or a half-baked strategy. They interpret and translate it for the masses. They integrate. You call, and they answer. You message, and they reply. They align. Coordinate. Connect. You throw them a brick, and they hand you back a wall. They analyze. Unify. Focus. Sharpen. Pay equal attention to the intern and the CEO. Filter out the noise and let the signal through.

Every engineering organization worth its name needs—at the very least—one tightrope walker.

Tightrope walkers move comfortably through the cracks. Fill the gaps. Play several moves ahead. Anticipate. Speed up when it's time to accelerate. Hit the brakes before the crash. Cut corners while looking both ways.

The tightrope walkers moderate. Bring opposing sides closer. Reconcile. Lower the temperature when things get heated. Listen. Think. Start and delegate. Mentor. Teach. Train.

The tightrope walkers finish what they start. Remember. Backtrack when they screw up. Apologize. They learn.

Deliver on time and in full. They don't run away from the hot mess. They ask for the ball.

Never let a tightrope walker go. They tie the room together.

27.7 The Technocrat

When Archibald Putt coined the term *technocrat* in his famous book, he was describing an interesting creature, a kind of survivor in the sea of nonsense.

In Putt's world, the technocrat is the engineer who learned the rules of the managerial game without fully succumbing to them. Not the converts who would sell their mothers once they are christened as managers, but a quietly competent practitioner who knows that technical truth is not enough to win battles in organizations where appearance often means more than performance. The technocrat is fluent in both dialects: the technically accurate language of engineering and the vague, slippery jargon of management.

Unlike the unremarkable bureaucrat—whose goal is to perpetuate the process itself for protection—the technocrat's aim is to keep the machine running *despite* the process. They understand bureaucracy because they know that, more than an obstacle to work, it is a practically unavoidable occurrence. They can't get rid of it, so they learn its patterns and steer through the gaps.

Putt's technocrat is also a realist about visibility and credit. They know that competence, left unadvertised, is indistinguishable from invisibility. However, technocrats resist the urge to become just another org chart acrobat. Their skill is in maintaining just enough visibility to avoid being ignored, while staying close enough to the technical core to remain valuable and do meaningful, concrete stuff.

The technocrat knows that meritocracy in complex organizations is a fairy tale, but this doesn't make them bitter. The technocrats' authority emerges from competence demonstrated, and it's noticed and remembered by others.

Putt observed that organizations tend to elevate those most skilled in the art of appearances, not those most capable of understanding reality. The bureaucrats might climb faster, but the technocrats shall last longer. When crises come—as they always do—it's the technocrat who patches the failing systems, the one who still remembers how stuff works. Next time a true crisis breaks out, check under the tables, and you will see all bureaucrats hiding down there.

In that sense, the technocrat is both an engineer and a bit of a sociologist, closely watching the weird rituals of management unfold while ensuring the machines stay running and products keep working.

27.8 The Anti-skilled

Agh, back to the messy ones. If you ever find yourself working with someone who seems incredibly skilled in doing the wrong thing, worry not, you are just working with an anti-skilled. Ok, maybe worry a bit.

According to the dictionary, a *skill* is a learned capacity of doing something competently; a developed aptitude or ability. Being skilled at something also means consistency: you are genuinely apt to do something if you can do it several times with more or less degrees of varying accuracy and still be successful in your task. If you're playing bowling, scoring a strike once could have been just pure luck. Scoring ten (and not necessarily all of them in a row) indicates that you know what you're doing. Getting skilled at something requires a good amount of trial and error, where failures are noted and fed into the loop to avoid repeating the subpar outcomes, thus increasing the probability of success.

Let's look at the opposite side of the spectrum, where the inept lies. If you screw up something once, well, it happens, and you could do better next time. Now, if you bomb something consistently, one after another opportunity you are given, that's also some sort of a skill. A somewhat odd skill, sure, but a skill at last. So, let's call it an anti-skill.

Think of bowling again: if the ball goes to the gutter once, twice, thrice, ok, I mean it can happen; it's not easy to get the stupid ball centered in that slippery lane. Now, if you throw 50 times and ALL of them meet the gutter and not a single pin goes down, that's a skill. It takes effort to be so consistently bad.

If you look around yourself at your workplace, you will find plenty of the anti-skilled kind. People who screw it up over and over by making wrong calls after wrong calls in an eternal chain of haphazard decision-making. They are anti-skilled specimens in action worth observing (from a safe distance).

You could think that one of the most fascinating parts of the anti-skilled is their unrelenting refusal to learn. Feedback, evidence, data, and failure appear to be shrugged off with the grace of someone convinced that the universe, not their decision-making, is the problem. But I disagree: the anti-skilled DO have a learning loop in action, only tuned with a minus sign added somewhere in the chain. They are continuously learning how to do it "better" every time, which means worse if seen from the skilled frame of reference. Those who absolutely master the craft of screwing things up are anti-geniuses. Anti-Einsteins of sorts, probably made of antimatter.

However, be careful if you ever find two or more anti-geniuses teaming together, as the sum will be greater than the individual contribution of the parts. When anti-skilled individuals collaborate, they create a multiplying effect, each validating the other's worst instincts.

The anti-skilled are skilled in screwing up through learned, reinforced, and continuously refined practice. And that makes them far more dangerous than the clueless, because they bring the full force of commitment and determination to every wrong turn they take. You can train incompetence out of someone who doesn't know better. You can't train it out of someone who's spent years perfecting it.

When you mix anti-skills with entrepreneurship, you get an anti organization. A company led by the anti-skilled that nobody wants to work for and that produces nothing. If you're an engineer trapped inside one, you'll know it shortly after starting there. Everything about it is absolutely wrong, as per (anti)designed.

27.9 The Significant Other

In almost every organization of sufficient size, there is a certain character that appears and needs no interview process, no technical assessment, no cultural fit evaluation; they only need kinship to automatically bypass all the heavy processes any other mortals would need to go through.

The Significant Other (SO hereafter) is a peculiar occurrence in the organizational landscape, particularly common in startups and small-to-medium enterprises, where the lines between personal and professional blur beyond ethical levels. In many cases, they occupy invented positions on the org chart, navigating through the muddy waters of unnatural authority and irrelevance, but with an attitude as if they naturally deserve it. The SO arrives suddenly, and their presence creates an immediate distortion in the social fabric of the organization.

The SO operates with a weird form of untouchability. Everyone else must justify their existence through work, meetings, and performance reviews; however, the SO's job security is guaranteed by factors that have nothing to do with competence. They can be spectacularly wrong, consistently incompetent, and universally disliked, but they persist because there is no real "process" for firing them; nobody knows how to do it, not even those who are related to them.

This creates a dynamic where the SO also feels compelled to prove their worth precisely because everyone knows they don't need to. They become hyper present, launching initiatives that no one asked for, reorganizing things that worked perfectly fine, and making themselves impossible to ignore, which only reinforces everyone's suspicion that they contribute nothing of value.

The balance sheet for significant others is practically always negative. In all fairness, there are truly talented SOs who face the challenge that no one will ever believe they earned their position. They could be brilliant, hardworking, and valuable (some I know are), but they'll forever be seen as walking nepotism. Every success will be attributed to privilege, and every promotion will be assumed to have been decided over dinner at home.

The SO also carries confidentiality dilemmas everywhere they go. They know things they shouldn't know, and even if they don't, everybody thinks they do. They operate under different rules than everyone else because, in the end, they literally sleep with the one who makes the rules.

The SO can poison feedback loops. Almost no one will tell the boss what everyone actually thinks about their significant other, as it could risk their careers. Thus, the SO never receives the honest feedback that might actually help them contribute more meaningfully to the cause.

The Significant Other ultimately represents a reminder that organizations, for all their talk of meritocracy and rewards, are still frequently run like royal courts.

27.10 The Single Point of Failure

Now it's time to talk about that one person whose departure could create an Armageddon, and yet no one is doing absolutely nothing to mitigate that risk.

The path of becoming a single point of failure is paved by organizational indolence, where the single points of failure grow in plain sight but so slowly that, by the time they have become a vital piece of the mechanism, it feels like it's too late to do anything.

Often, the single points of failure are highly capable, borderline brilliant people with somewhat "special" personalities, which doesn't make them the best mentors for passing their knowledge along, or the best hiring managers to hire those who could become clones. Many single points of failure are absolutely aware of what they are and may weaponize this to guarantee themselves job security or simply feel the thrill of being indispensable.

Once, I worked with an engineer who was doing firmware for industrial equipment, and this guy was the only employee doing firmware in a whole company of 30–40 people. This person—he was also one of the founders—would only handle binary code for us. He would keep all source code in his own computer, and his office would be a protected castle he kept closed to practically everybody except a few selected people who were part of his small, inner circle of trust, which I wasn't part of. I would understand someone exploiting their role of single point of failure for job security when you're just an employee with a contract, but this person was one of the owners, so you would imagine their job security was not precisely at stake. Still, when there was a firmware issue at the factory floor or we needed the latest version of something, this guy was, of course, the only person who could do it, so the factory people would call him, he would step out of his office and come down the stairs walking like a rockstar, loving every single second of attention. One could really see why he would refuse time and again to mentor someone else for the job. He absolutely loved being a single point of failure. There are many such cases.

However, what makes the single point of failure role so sexy due to the high attention and the aura of dependence, it also makes it its biggest curse. Being a single point of failure means lacking all freedom because, when bad things happen, you are the only one with the knowledge to potentially sort it out, and this can happen during weekends, nights, Christmas, and whatnot. Besides the brief moments of glamour, being a single point of failure is quite a miserable place to be in an organization, something that the single points of failure themselves can't even see because they are blinded by the tree of self-importance, missing the forest completely and holding an entire organization hostage under the belief that being a bottleneck is the ultimate flattery.

27.11 The Delusional

Some people seem to go on life as if they were connected to VR headsets 24/7. Whether a mechanism of defense to boost their self-esteem or early signals of mental decline, they choose to cut ties with reality and make it very weird for those who are not part of the delusion.

Circa August 2016, and I am North of Edwards Air Force Base in the Mojave Desert in the summer with nearly 50 degrees Celsius, about to witness the launch of a rocket made by an early-stage launcher startup. The "rocket" is, in fact, a little sounding rocket that looks quite rudimentary. The person in charge of this test launch—a man in his mid-fifties with the looks of a mad scientist who probably votes Republican—looks tough and absolutely distrusts these two foreigners—me and my colleague—coming from Finland, who are there to attach some computer at the tip of their tiny rocket for testing purposes. The man asks us a lot of questions about our origins, and our badges show bright, red-colored dots, indicating that no one is allowed to share any information with us whatsoever due to ITAR restrictions. While he's still interrogating us, I can't help but look at the rocket again, and the whole situation feels ridiculous. Every now and then, I pinch myself just to check if this isn't a hallucination from a heat stroke I might have had, or maybe it was one of the plethora of deadly animals that live there that bit me, or something.

As the thing slowly develops, I can see how all actors involved in the test launch are acting like they are about to launch Saturn V or something. Nothing that I see actually matches the amateur level of the whole operation.

At some point, and after the obligatory and dramatic countdown by the boss that I can listen to thanks to loudspeakers, the little rocket lifts some unremarkable 200–300 m maximum, coming down with a cute little parachute, leaving a small cloud of dust as it hits the ground. Our computer couldn't gather a single point of data, considering the flight was just too short, and our sample rate was too slow, expecting a longer flight time.

Once the bruised rocket is brought back to the hangar (read, a tent) carried by two men on their shoulders, the boss and the rest of the team seem relieved and happy with the result, and everybody now relaxes with beers and snacks as they pat each other's shoulders and check the videos captured by drones. They talk about "money shots", they are all ecstatic.

And as I am just sitting there, sweating like Richard Nixon, I can't help but think I am witnessing what seems a collective delusion. I can see how the boss is probably perceiving—read, hallucinating—the rocket as huge, spitting oxygen in furious vents, stunning, and fully erect (always talking about the rocket). And how he sees himself as the mighty launch director of that majestic flying machine, I bet, mentally situating himself in the Cold War era where he's firing his creation to show those commies who's the boss, wishing to be able to ride it, Dr Strangelove style. Of course, all absolutely disconnected from the painfully modest reality of something that is just slightly bigger than a firework, struggling to get off the ground.

Now, circa maybe 2012 or 2013, another mad scientist boss is fronting a launch, and I am following this live on YouTube. The launch, this time, is of a nanosatellite, which is apparently the first satellite ever launched by a certain country. It's a Cubesat; Cubesats are launched by agencies, companies, and universities alike. They're tiny, cute, and useful due to their cost-effectiveness and low complexity. However, this person is treating the launch like it's a mission to kill an asteroid that is about to end life on Earth. His nanosatellite is epically named after some mythological creature and is being launched somewhere in China. In the live video, you can see the launcher staff busy minding their business, probably taking care of the rocket telemetry and the primary payload. Now, this mad lad (who is also wearing Star Trek-like clothes) is making it look like the whole operation, the entire staff of tens or even hundreds of flight engineers, is entirely dedicated to monitoring the status of his tiny satellite.

And I am there, sitting in silence; a distant, nameless spectator in absolute awe, half amused by the bizarreness of the whole thing, half alarmed at how someone can even think this childish facade could convince anyone with a slightly trained eye.

On some other occasion, I was part of a project where we were working with a small, early-stage company that had disproportionately big dreams. Go big or go home, they say, I guess, and that's fine, but these folks had crossed a line between ambition and collective delusion. They had bombastic philosophical slogans on their website about changing society, this and that, all while having no pockets to even remotely achieve what they wanted to achieve. I am aware that bluffing to incrementally increase your runway as you mature your product is a common strategy in the startup world, and there's no harm in it, but these people were not doing that: they were insistently shooting for the stars while equipped with a water gun. We, mere suppliers, were carefully and politely trying to nudge them back into sanity, trying to find a more incremental approach aligned with their realistic possibilities, and yes, it felt weird to step into their strategy, but they were too detached from reality not to do anything. All we got were angry letters from their CEO lecturing us on how small we were in our thinking and how far we were from their wet dreams of disrupting everything. It was pure madness, and I am telling only 10% of the story. Anyway, it was hardly a surprise when, sometime later, they nearly went bankrupt, the CEO was sacked, and they pivoted into something else, something less transcendental and more practical.

Of course, another way of reading this would be that I've been the delusional one all along, suffering from some perception disorder that makes me see everything shittier, crappier than it is. And that could be true.

However, do note that I am not using the word "delusional" in an offensive way. As per the dictionary, a delusional person is characterized by holding false beliefs or judgments about external reality that are held despite overwhelming evidence to the contrary, and this is precisely what I am talking about: when evidence shows you something that doesn't align with what you want or expect, you choose to ignore the evidence and interpret reality your own way.

It appears as if, when someone's brittle ego and/or dented personality become entangled with a project or activity, it becomes an extension of their identity. The more fragile the reality, the more elaborate the stunt required to compensate for it. However, these engineering delusions can, as we saw, reach a weird collective status, where a group of reasonably educated professional adults propel each other into believing something that is just not there, failing to see things as they are, because things don't look the way they would like them to look.

As I sat there semi-dehydrated in that tent in the Mojave Desert, questioning my life choices, or watching that overacted nanosatellite launch livestream, or in a meeting with a deranged CEO, I realized that, in all cases, what I was seeing wasn't engineering anymore, but more of a theatrical display; the human reflex to decorate a modest achievement into something epic enough to fill the uncomfortable void between ambition, and capability, expectations and reality.

And since we are on the topic of reality distortion, this couldn't be a better moment to jump to the next chapter, where we keep on talking about twisting reality as if it's some Play-Doh, this time when people intentionally distort others' reality using their position of power to achieve whatever they want, all in the name of "inspiration" and objectives.

Chapter 28
Unrealistic Goals and Reality Distortion Fields

Abstract Managers will try to distort reality to make you meet their unrealistic goals. In multidisciplinary engineering design, goals cannot be defined in isolation at the lowest levels of the hierarchy of the system under design. They must all align with the system life cycle and the architecture. There is a tedious, large bibliography body written about goal setting, always with a new way to skin the same old cat. Besides the usual babble, goals must always be tied to (undistorted) reality, to the underlying work breakdown, and more fundamentally, adaptive with respect to the evolving context.

Keywords Goals · Objectives · Project management · Management by objectives · Reality distortion field · Leadership

Among the many things Steve Jobs is credited with doing, he is also attributed with deploying his infamous "Reality Distortion Field" (RDF) to influence teams to accomplish what seemed impossible. This term, originally coined by Bud Tribble—some Apple dude—encapsulates the quasi-mystic effect Mr. Jobs wielded over his employees, bending perceptions of what was realistic or achievable. While the distortion field might seem unique to Jobs, it's a pretty common tool that many managers use, and it's a method that is both over-romanticized and dangerous.

In fact, the reality distortion field is a glorified synonym for gaslighting. Most books depict Steve Jobs' ability to subdue his subordinates as "inspiration". In reality, Jobs' capability to "inspire" his team to believe in impossible deadlines and implausible product features wasn't entirely inspirational mastery or high intellect but more like fact-twisting sprinkled with some good old authority abuse. Authority figures have it particularly easy at shaping belief systems and behavior, as Milgram's famous obedience experiments demonstrate how individuals are predisposed to comply with authority figures, even when their directives defy logic or even morality.[1] Hierarchical presence and an overwhelming confidence aren't necessarily signs of genius but rather simplistic mechanisms aimed directly at the team's critical faculties to

[1] Note that there is quite some debate about the interpretation of Milgram's experiments.

I. Chechile, *Engineering is Broken*, https://doi.org/10.1007/978-3-032-18169-5_28

coerce them into complicity. The feeling of your job being at stake if you contradict your boss's latest whim is enough deterrent to acquiesce and play along, even if it's the most stupid idea you've ever heard.

The human cost of this tactic is quite high. Prolonged exposure to unrealistic expectations erodes mental health, job satisfaction, and even physical well-being. Far from being signs of a successful team, they are clear symptoms of an organizational breakdown, coated by a manager's pathological need to appear infallible and visionary.

Some say some reality distortion is necessary for small organizations like startups. That you must aggrandize things to appear bigger than you are to appeal to investors, customers, and the like. The old bluffing again; we spoke about this. That distorting reality, when done for "good reasons", isn't bad. One can understand a certain degree of reality distortion towards the outside world. "Fake it till you make it", right? I suppose. However, what do you gain by turning reality into a mirage with your own people, with those you are supposed to build trust with?

Managers who employ these tactics are closer to liars than leaders. By distorting reality, they deprive smart employees of their agency, their critical thinking, and their ability to make informed decisions.

The Reality Distortion Field is a fancy way of saying manipulation, practiced by those who are too lazy or incompetent to lead ethically and according to reality, and too self-absorbed to acknowledge their flaws. True leadership is rooted in transparency, empathy, and realistic goal setting. Bend it as much as you want, but reality, after all, has a way of reasserting itself with merciless precision.

28.1 Setting and Adjusting Goals

In the late 1960s, the Ford Motor Company was losing market share to foreign competitors that were selling small, more fuel-efficient cars. Then-CEO Lee Iacocca announced a challenging goal: producing a new car that would be "under 2000 pounds and under $2,000" and would be available for purchase in 1970. This goal, coupled with an unrealistic deadline, kicked off a known behavior: start cutting corners to obsessively meet the goal that's been set. For the Pinto, it meant that many levels of management signed off on unperformed design rules checks to expedite the development of the car. One omitted aspect concerned the fuel tank, which was located behind the real axle in less than 25 cm of crush space, way closer than what most designs of the time used. The rest is history: A slight rear-ending would turn a Ford Pinto into a fireball.

In multidisciplinary engineering design efforts, goals cannot be defined in isolation at the lowest levels of the hierarchy of the system under design. They must all align with its life cycle. For example, if you are developing an aircraft, you cannot assign goals to the software team that only concern that team and some other goals to, say, the wiring team that only concern them, and expect that it will go well. It will not go well. What ultimately matters is how the aircraft—the system—will mature as a whole from life cycle stage to stage, so it is way better to set the goals "systemically". This sounds stupidly obvious, and yet many go and set goals per domain that don't tie to the goals of other domains.

Then there's the problem of unnecessary granularity when setting goals, which has become an issue of our times. Goals can perfectly read and sound coarse as long as they conveniently capture the work of relevant domains into a single, meaningful set point. At the same time, goals cannot stay rigidly fixed while everything else is moving and there are good reasons for that movement. Of course, if goals change too much and too frequently, that can be a problem, but reasonable goal adjustment is a sign of maturity from leadership, not a weakness.

Another significant challenge is that goals might create tunnel vision and inattentional blindness. In a well-known experiment, Simons and Chabris studied this effect by asking participants to watch a video in which two groups of players pass basketballs. One group wears white shirts; the other group wears dark shirts. Given the task of counting basketball passes among people wearing only white shirts, people unconsciously block out the black-shirted individuals. As a result of this narrow focus, most participants fail to notice when a person wearing a black gorilla suit saunters into the middle of the screen, pounds his chest, and walks off-screen. Intense concentration on the task causes people to overlook a striking element of their visual world. People might miss the gorilla if pushed too hard to achieve a particular goal. Think of the Pinto again.

Furthermore, goal enforcing and tracking eventually becomes a full-time job no one wants to do. The problem with this is that goals start to rot as everything does, and as soon as the team perceives the smell, goals stop serving their purpose, as the idea that no one cares anymore radiates in all directions.

Finally, the laws of physics are the ultimate limit for engineering goals. In short, no objective can ask for bending or breaking universal laws. Remember the Vasa again, and how the engineers chose to prioritize the King's whims over Archimedes' principle, and how that went down. I heard once from a wise veteran engineer: "I quit my job when my boss asked me to break physics' laws".

Ultimately, management by objectives is as good as the objectives.

There is a lot of boring bibliography written about goal setting, always with a new way to skin the same old cat—call it OKRs, MBO, or whatever acronym floats your boat—appearing every half decade or so, with the subsequent multitude of 350-page books thrown into the shelves. Books that will all basically parrot the same speech, preaching on how goals should be measurable, attainable, relevant, and specific. Besides the babble, goals must always be connected to reality and, more fundamentally, adaptive with respect to the context.

Football is often a great source of managerial wisdom. If only managers would watch more football!

On the 28th of March 1999, a match took place in the second division of Argentine football between Ferrocarril Oeste and Belgrano de Córdoba. It was a match of struggling teams; both were fighting to avoid relegation to the third division. There was a lot at stake. Ferrocarril Oeste had not scored a single goal in the tournament so far, and that's when the supporters of Ferro fielded a flag containing a request—which has ever since become mythical— so clear, concise, realistic, attainable, and measurable that it would have been the envy of many MBAs out there:

—Score a goal.

Chapter 29
Residual Taylorism in Engineering Design

Abstract The act of designing a system is different than the act of manufacturing it. Engineering design, being an activity obviously linked to fabrication and production, is dragged into management methods similar to those used on the factory floor. The ghost of Frederick Taylor hovers in design tables, dashboards, and Gantt charts, where it has no business chasing outputs with meaningless metrics. This residual Taylorism, born in the domain of conveyor belts and stopwatches, treats engineering creative work as deterministic and linear, when in reality it is messy, iterative, and full of uncertainty. From this logic springs pseudo-scientific indicators such as "efficiency" or the more obscure "productivity", at times used interchangeably, that create the illusion of control while measuring nothing in reality.

Keywords Taylorism · Engineering Design · Process control · Productivity · Efficiency · Metrics

We love to blur the boundary between the design process that ideates and develops the system and the manufacturing process that creates instances of said system.

Almost two decades and many kilograms ago, I was a young engineer full of dreams working at a family-owned small company where I was part of the R&D team in charge of designing certain industrial equipment.

The R&D department's office was located inside the factory where said equipment was manufactured. "Office" is perhaps too much of a word: it was a brittle mezzanine right above the factory floor, so I could hear the drilling, smell the welding, and breathe the heavy metals from the fumes stemming from the reflow oven pipes.

Because we were right inside the factory, the R&D department followed the same rules and customs as the factory workers, including scheduled 10-minute breaks in the morning and the afternoon that were signaled by the sound of a horn. Fun times.

In the factory, there was a group of workers in charge of manually inserting through-hole components in the large printed circuit boards (PCBs) before the boards would be sent for wave-soldering and reflow. Each PCB would have thousands of components of the same kind, so it was an excruciatingly manual and repetitive task for those involved. The bosses had a system to gauge how many components per

I. Chechile, *Engineering is Broken*, https://doi.org/10.1007/978-3-032-18169-5_29

hour a worker could insert, and such a metric was directly impacting their pay: they would get bonuses if they showed an improvement in their performance.

At some point, our boss at R&D was requested by the company owners to come up with similar metrics for his team. That meant us. Who was "us"? A collection of nerdy, young programmers, industrial designers, and electrical engineers. In short, my boss had the task of finding a way to gauge and measure, factory-style, how all these engineers performed.

And the man struggled.

There were attempts to measure lines of code per day. Another attempt was to measure "chair hours". There were spreadsheets to be filled. He started crafting intricate plots in his office. Soon, he and the company owners realized the pointlessness of the fabricated metrics, as they were mostly random noise unable to describe or predict anything related to how efficiently engineers were doing their jobs. There was this brilliant guy in the team who would slack off for days and then would pull the most intricate code in a matter of a few hours (we used to call him "Dirac" because he worked in impulses). There was another guy who was not the most hard-working person ever (chair hours were not his thing), although he could put together very complex technical thoughts together with a facility that we all envied. That guy was doing really badly from the new metrics' perspective, although he was an essential part of the team's success.

In the end, our boss dropped the idea of measuring our engineering work because he realized those weren't the droids he was looking for. It was, perhaps, not the highest performing R&D team ever, that's for sure, but it was decent in the sense of creating new designs and products at a good pace to keep the company's leadership in the market. Isn't that precisely what R&D teams are for?

The Industrial Revolution disseminated the idea that work, regardless of its nature, had an inescapable time dependency. From this mindset, the tasks related to assembling a car or designing a car were equally subject to time measurements, and conclusions were drawn from such measurements to "improve" them if the numbers were not satisfactory.

Then, as the early twentieth century unfolded, witnessing the end of the second industrial revolution, Frederick Winslow Taylor made an infamous appearance.

Taylor worked for many years in heavy manufacturing, where he climbed his way up the job ladder: laborer, shop clerk, gang boss, foreman, and finally chief engineer. During that time, Taylor observed everything around him, and what he saw disturbed him. The way jobs were done varied wildly from person to person. Some people were quite proficient. Others weren't. So, he quit his job and devoted the remainder of his life to tackle inefficiency in some systematic way. And in doing so, he gave birth to scientific management, also known as Taylorism.

At the core of Taylor's philosophy is a belief that management had, in the words of Taylor himself: *"... laws as exact and clearly defined as the fundamental principles of engineering..."*. It's a lovely irony that our friend used engineering as a yardstick for his observations, isn't it?

See an extract from the introduction section of his "*The Principles of Scientific Management*":

"This paper has been written:

First, To point out, the great loss which the whole country is suffering through inefficiency in almost all of our daily acts.*.

Second. To try to convince the reader that the remedy for this inefficiency lies in systematic management, rather than in searching for some unusual or extraordinary man.*.

Third, To prove that the best management is a true science, resting upon clearly defined laws, rules, and principles, as a foundation. And further to show that the fundamental principles of scientific management are applicable to all kinds of human activities, from our simplest individual acts to the work of our great corporations, which call for the most elaborate cooperation".

And then he goes:

"...the greatest prosperity can exist only as the result of the greatest possible productivity of the men and machines of the establishment—that is, when each man and each machine are turning out the largest possible output; because unless your men and your machines are daily turning out more work than others around you, it is clear that competition will prevent your paying higher wages to your workmen than are paid to those of your competitor. And what is true as to the possibility of paying high wages in the case of two companies competing close beside one another is also true as to whole districts of the country and even as to nations which are in competition. In a word, that maximum prosperity can exist only as the result of maximum productivity".

This is some true hardcore Taylorism compressed in just a few paragraphs. As you can see, one of Taylor's main worries was how to avoid depending on the "unusual or extraordinary man", so he created his systematic management as a protection for that. Taylorism is a philosophy that averages down work for everyone by design, as it makes sure nothing depends on a few great performers but on many average ones. And this right here is something that doesn't apply to engineering design, where it always pays off to find or develop people to become "unusual and extraordinary" rather than to train a myriad of average designers. You can probably assemble a car with many average workers by telling them exactly what to do, what bolt to tighten, and with what tool to be picked from where, but you can't create and design great things with many average engineers. To create great stuff, you need the "unusual and extraordinary" people that Taylor was so worried about.

And yes, the man truly believed humans and machines could be treated alike. Taylor's was one of many other attempts to apply mechanistic methods to analyze workers using rather simplistic mindsets, aiming to predict and mathematically model human behavior. A school of thought that finds its roots in the somewhat cute (and damaging) "envy of physics" and the scientific method. If the behavior of stars and planets can be accurately measured and predicted with a set of elegant equations, then any system's behavior should be able to be captured similarly, including people gathered together for work.

It is unsettling that much of today's work is still affected by management ideas stemming from ages ago, and in particular from Taylor's work.

Additionally, in anything social—like engineering work is—the symmetry between measurement, prediction, and explanation is broken because the future in this kind of activity is genuinely uncertain, and therefore cannot be assessed or predicted with the same degree of certainty as it can be explained in retrospect. When it comes to people and what they do, we have as yet only a very imperfect ability to tell what has happened in our managerial "experiments", much less to ensure their reproducibility or measurability. Some authors, like the Nobel-prize economist Friedrich von Hayek—a debatable figure, but a man of straight points—labeled this as *scientism*: an approach to brute force natural science methods into distant areas, in this case, the social sciences.

What is more, natural sciences put the observer as an external entity from the phenomenon to be observed, detaching them completely. In anything social, like work, thinking is part of the reality that the observer has to think about, which makes the relationship circular. The complexity of the world exceeds our capacity to comprehend it, so we are obliged to resort to various methods of simplification: generalizations, metaphors, decision rules, just to mention a few. Fallible human beings are not merely "scientific observers" but active participants in the system themselves. Even more, such imperfect views can influence the situation to which they relate through the actions of the participants. Whenever we find ourselves on the task of measuring a social arrangement like a team, we are part of the analysis, and our biased perceptions can influence our decisions to alter the scenario in a way that reinforces our own beliefs. Taylor, or any other "observer" for that matter, could never prescribe or mandate what drives productivity or efficiency because they are part of the whole thing, therefore part of the problem.

Taylor finishes his text by trying to define what "scientific management" is:

It is no single element, but rather this whole combination, that constitutes scientific management, which may be summarized as:

Science, not rule of thumb.

Harmony, not discord.

Cooperation, not individualism.

Maximum output, in place of restricted output.

The development of each man to his greatest efficiency and prosperity.

The exact link between prosperity and productivity is never elaborated in his book, but one can be sure he meant prosperity of the shareholders, not the workers' prosperity. Please be more productive so someone else can get richer.

All in all, productivity is ultimately a "global" attribute of the organization. The most productive engineering department ever, coupled with an unproductive sales department, still equals an unproductive organization. The belief that "if everyone is more productive, the whole becomes more productive" is false because work is compartmentalized, and productivity is also a function of the coordination between the different groups and departments. We have seen too many times highly productive engineering teams with their outputs totally nullified by incredibly unproductive management.

Efficiency—understood as how wisely we use resources to create value—is ultimately about working smarter. Ideally, the most efficient way of working is doing absolutely nothing. That means devising a way of creating economic value with the least amount of involvement and consumption of resources possible, so we can go and enjoy life with friends and family. More efficiency can be achieved with training, with good tools, with unobstructive processes, with sensible automation, and with managers letting people do without interrupting them every five minutes for minutiae.

29.1 The Ethereal "Processes"

In the engineering practice, there's this mantra that goes that if you can't describe what you are doing in a process, then you don't know what you're doing. Some others preach that "All work occurs in a system of interconnected processes". This is also factory-floor thinking, which has no business in engineering creative work.

The problem is that creative and uncertainty-rich activities—very present in our craft, in particular at the early stages of an engineering project—cannot be captured by the historical concept of a process, and that does not mean we do not know what we are doing.

But what is a process, though? The dictionary says:

A process is a collection of related tasks executed by one or more team members, whose execution produces a service or product that serves a business objective.

In industrial environments, capturing what a "process" is can be rather straightforward: assembling a watch, etching a pattern on a semiconductor wafer,[1] drilling a hole, or filling cans with a fluid. To monitor, analyze and potentially improve any process, it is necessary first of all to identify what the process's boundaries are and also what its inputs and outputs are. Many processes are "discoverable" and can be described with tangible procedures, modeled and studied with statistics, but when it comes to creative engineering practice, Gaussian curves just don't work anymore. Optimizing an algorithm, debugging, running trade-offs; in many situations around our craft, it can be difficult to define the "process" because things do not show easily identifiable inputs and outputs as industrial processes do. The entire science of process control that many precision industries rely on is useless when the activity is based on creativity, problem-solving, and trial and error.

However, the engineering managers who have let the ghost of Taylor and Toyoda conquer their souls can't help thinking that the entire engineering craft is like making a watch or manufacturing shoes, so off they go, forcing "processes" to appear when the process concept is rather inapplicable, all fueled by the thinking that human creative behavior can be captured in flowcharts as if we were tokens in a Petri net.

When faced with sloppy practices and bad outcomes, bosses will happily flex that "a process will be put in place" and even things conceived for the factory floor like Lean Six Sigma, and other fancy terms will be randomly thrown into the mixer as miraculous solutions. People will attend trainings and certify in this and that. I have

[1] By no means do I mean "straightforward" in the sense of "easy". This is an extremely complicated process, but it's an industrial process in which inputs and outputs are rather identifiable

nothing against Six Sigma, but reducing the variance of a manufacturing process is one thing (and it works wonders), but good luck reducing the variance of a creative process where variance is far from having a "close form".

Again, at the risk of being flagged as "un-rigorous" by all the little Taylors lurking out there, I need to promptly state that, just like it happens with work packages and requirements, using processes the right way for the right thing makes sense. My main argument is that not every activity done by engineering professionals can be captured in an explicit sheet of paper, a procedure, an equation, or a checklist. Furthermore, the earlier in the development process, the harder to devise processes because determinism it at the lowest while uncertainty and creativity are at their peak.

A process makes sense to be made explicit when the tasks involved are truly sequential and well-defined, where inputs and outputs are identifiable, for instance when I'm operating or assembling a machine: materials here, outputs there, push here, check that, watch out for X or Y because if I don't follow the list, I die (or someone else does). In that case, give me checklists, procedures, statistics, flowcharts, give me tools that help me focus and avoid overloading my brain. Otherwise, we're talking about engineering creativity, design work. In short, high-variance stuff; the process to design a Boeing 747 is not the same as assembling it, not the same as flying it.

Engineers are not symbols in a diagram.

29.2 Out-of-Context Efficiency

I start from the punchline: The blind pursuit of efficiency and optimization without assessing why those supposedly "suboptimal" things were there in the first place can introduce risks that most of the time invalidate any marginal gains achieved.

On May 25, 1979, American Airlines Flight 191 crashed shortly after takeoff from Chicago O'Hare International Airport, killing all 271 people aboard and two on the ground. The immediate cause was the separation of the left engine and pylon assembly from the wing during takeoff.[2] While this appeared to be a mechanical failure, the accident's root cause traced back to a maintenance procedure that had been modified to improve efficiency.

The aircraft manufacturer, McDonnell Douglas, had specified a procedure for removing and replacing the engine and pylon assembly (the structural component that connects an engine to the wing or fuselage) that required separating these components from each other before removal. This process, while time-consuming, minimized stress on the critical mounting points. American Airlines, seeking to reduce aircraft downtime and labor costs, adopted an alternative procedure that removed the engine and pylon as a single unit using a forklift. This method reduced maintenance time from approximately 200 worker-hours to just over 40 hours per aircraft, a seemingly impressive efficiency gain.

[2] At the time of writing, a new accident shows similar characteristics to Flight 191, although it's still soon to say whether it's the same root cause or not

However, this streamlined procedure introduced substantial risks. The combined weight of the engine and pylon assembly, coupled with the difficulty of achieving precise alignment during reinstallation, created paths for structural damage to the mounting hardware. Over multiple maintenance cycles, this process caused fatigue cracks in the pylon's aft bulkhead attachment point, damage that went undetected until it resulted in catastrophic structural failure.

This case illustrates several critical points in any efficiency-driven decision-making. First, the modification sought to optimize for a narrow metric—maintenance time—while not considering the broader system implications. The procedure reduced costs while creating hidden risks that accumulated over time. Second, the efficiency gains were concrete and measurable, while the impact on safety was too gradual and difficult to detect. Third, the organization lacked sufficient feedback mechanisms to detect the emerging risk pattern before it manifested in disaster. Yes, I am aware there is some survivorship bias at play here: there probably are other cases where similar changes in procedures were done in the search for increasing efficiency which didn't end up in disaster, and some efficiency was perhaps gained, but when safety is involved, only one accident is enough to understand it's not worth it.

The Flight 191 tragedy shows that procedures created by design engineers may often contain "inefficiencies" that serve crucial but non-obvious functions. When managers or operators alter these processes without fully understanding their underlying logic or original design drivers, they may inadvertently remove critical safeguards. The "waste" they think they're eliminating may actually be margin for error, or protection against rare but catastrophic failure modes. This one goes to all the Taylorists out there: Not all inefficiencies are wasteful; some represent essential systemic features, including safety measures. Maintenance is an important lifecycle in systems engineering, but it can be temporally so far away from the design stage that important information might fade and be lost in the journey.

The pressure to do more with less can lead engineering organizations to screw up in pursuit of the always-glorified "leanness". Not everything that is perceived as waste is. Blind pursuit to reduce waste can lead to disaster. Seeking efficiency must be done by a deep understanding of the complexity of the systems we deal with and awareness that optimization in one dimension may create vulnerabilities in others.

Chapter 30
And Yet It Works

Abstract We have developed an entire science of failure, but very little literature exists on why systems work. While failures are dissected and catalogued through post-mortems and crash reports, success remains largely unanalyzed. Given the imperfection of everything we build, one can't help but wonder why they can work at all. Every engineered artifact—from aircraft and reactors to the electronics on a lab bench—operates under a hidden choreography of flaws, variances, and latent defects. This chapter explores the paradox of reliable function emerging from unreliable parts, the hidden probabilities that keep our machines alive, and the interplay between complexity, reliability, and human error.

Keywords System reliability · Complexity · Failure analysis · Near misses · Human error · Probabilistic systems · Stochastic processes · Emergent behavior · Technological fragility

There are many sources and literature about how systems fail. There are crash reports, detailed forensic analyses of famous disasters involving complex systems such as nuclear reactors, ships, rockets, and aircraft. There are millions of words written about why failures happen and how to prevent them. There is, without a doubt, a science of failure.

But there isn't much written, though, on why systems work. And yes, I am aware of the initial absurdity of the question. It is like asking medical science to write about why people are healthy. Granted, it makes more sense to research about diseases—the off-nominal situations—and not the normal scenarios.

Unlike living organisms, the technical artifacts we create are not the result of selfless evolution but, on the contrary, subject to market forces, tolerances, production imperfections, and many other factors out of our control. With so many odds against, asking why artificial things work is not so absurd.

When I got my first job back in 2002 as an all-rounder in a now-extinct small company, I was fortunate to witness the development of this electronic system the team was working on. As a young engineer wannabe, I saw this device going from a scribble on a whiteboard, all the way to block diagrams, schematic drawings,

I. Chechile, *Engineering is Broken*, https://doi.org/10.1007/978-3-032-18169-5_30

embedded software, prototypes and—finally—the real thing sitting on a table. After the team fixed some initial mistakes here and there, the board firmware booted up, and some LEDs started blinking. At that moment, a strange feeling invaded me. Just as if something magical had just happened. Sure, the board was not a nuclear reactor in terms of complexity, but it had a fair amount of density, with parts and components coming from different vendors across the world, many different interfaces and protocols, and this sequence of mysterious, human-defined instructions—software—orchestrating the behavior of the whole thing. I saw the heated meetings where engineers discussed features. I saw the budget expanding and shrinking. I saw the delays in deliveries and the manufacturing artifacts in the components sitting on the board; some looked a bit crooked, and some solder looked cold. And yet it worked.

That simple question stayed with me for more than 20 years: how the hell did that work? The same question assaults me every time I am in front of any complex system that involves a large number of components. How on earth do systems with hundreds, thousands, or millions of parts manage to turn on and run? I feel that, as engineers, we get too used to the fact that things work, and we do not stop to appreciate how the probabilities are mysteriously playing on our side. I somehow refuse to get unemotionally accustomed to seeing complex things work, although I can surely live with it, but not without constantly thinking about the underlying probabilities playing in my favor.

I know why systems are *supposed* to work in the design domain, where everything is well-behaved. Where we control most of the variables, and where everything abides by deterministic physics and circuit theory laws. However, complex systems working in production, in the field, are a different story. Think about it. Myriads of components, all with their own life span, their own variances, their own bathtub curves, their own intrinsic design imperfections, tolerances, bugs, and noises. A Boeing 737—with over 10,000 units produced worldwide—has an average of 600,000 (six hundred thousand) parts; the Airbus A380 has nearly four million. If we can assume no perfect system exists in the world—which is an undeniable fact—we can also say the same for any subsystem, sub-subsystems, all the way down to the most elementary part. Nothing is perfect. So, how can reliable function emerge from a collection of myriad imperfect elements hooked together? The surprise is not that there are so many accidents. The surprise is that there are so few, considering that the world is not well-behaved.

When we talk about why systems do or do not work, we can't really leave human errors out of the equation. Although we live in an age of autonomous systems, we humans remain a critical piece of the puzzle. Therefore, human errors are very relevant when it comes to analyzing complex systems' failure. For example, in the early days of flight, approximately 80% of accidents were caused by the machine, and 20%were caused by human error. Today, that statistic has reversed. Approximately 80% of airplane accidents are due to human error, and 20% are due to machine failures.[1] These statistics leave a few questions:

[1] Boeing Aero magazine, Q2, 2007.

- Have we humans become sloppier operators in time?
- Have systems become more complicated to operate?
- Have systems become more intrinsically reliable?

Nothing really indicates we have been worsening as human operators, but systems have indeed become more complicated to operate. When it comes to reliability, there is an undeniable increase in intrinsic reliability with better materials, better tools, and better manufacturing.

No system in the history of humanity has been made perfect. There is always a non-zero number of flaws on every system ever made, or to be made. Every single plane you and I have boarded or will ever board has a non-zero number of software bugs on board. Same for every nuclear reactor that is currently operating around the world. The question is, how relevant are those defects to bring the whole system down at once? All the systems we use, we fly on and operate, show errors at almost any given time. Every existing vehicle out there on every road is having at this precise moment checksum errors, data corruptions, bit flips, material cracks, and bolts getting infinitesimally loose every millisecond. Be it planes, cars, nuclear reactors, or hair dryers. Some of those events may never get reported, and perhaps even noticed or logged, and may be obliviously corrected by routine maintenance.

A usual argument is that complexity is the enemy of reliability. But is it? A Bleriot XI (practically a flying bicycle, go see a picture of it) is abysmally simpler than a cargo Boeing 767, both carrying a crew of 2. It's easy to see which one is more complex. However, which one is more reliable? Which one would you choose for taking a flight across the Atlantic on a foggy night[2]?

Revisiting the question of why we don't see more complex system failures, I would rephrase this a bit differently. Systems are rife with accidents. Most of the time, very small ones. Systems coexist with imperfections and errors. They continuously sustain an organic number of internal defects and "benign" off-nominal behaviors that they manage to live with. As we equip the artifacts we create with more parts and components, the path to systemic, global failure gets longer, more intricate, and thus less probable.

Is it that things work because they are complex, then? Is complexity a valid protection for failure? Not really, and here comes the third dimension, which compensates for anyone thinking that over-engineering to create longer, more intricate paths to failure is the key to protecting themselves from disaster. More complexity exerts more pressure on whoever has to deal with the more states and transitions complexity brings, whether a human operator or an algorithm. Any potential reliability increase brought by complexity is compensated for by an increased risk of operational error. You could get to fly the Bleriot XI or the Wright Flyer after a few tries, but flying an airliner is something else. Again, which one is more reliable? The answer depends on who drives.

---

[2] Interestingly, Charles Lindbergh deliberately chose a single-engine design for the Spirit of St. Louis to maximize range, minimize weight, and reduce complexity. In 1927, one reliable engine was more trustworthy than two mediocre ones and less drag and fuel burn meant a better chance of reaching Paris.

Ok, so why do systems work, then? It largely depends on the zoom level when you look. From an external observer, systems work because their internal faults are not aligned to "win" the disaster lottery. Under the magnifier, systems are failing all the time.

30.1 Near Misses

Some engineers seem surprisingly unaware of the concept of a near miss. Some others are aware but love to dress them as epic wins, brushing off the fact that the situation almost ended in disaster. A near miss refers to a situation in which an event or incident almost occurs but does not result in any harm, injury, or damage. It can be described as a "close call" or a narrowly averted accident. In other words, a near miss is a situation where the potential for harm exists, but by pure chance or timely intervention, the negative outcome is prevented.

A group of people gathers at a distance from an old clothing factory in the Czech Republic to watch it get demolished. Just as the charges are detonated, a stray chunk of masonry flies from the collapsing building, narrowly missing the group of boom-gawkers. When watching the slow-motion video,[3] it is obvious that, had the chunk of concrete hit the onlookers, it would have instantly killed them. Luckily for these folks, it missed. Barely.

On the flip side, Mike Edwards, 62, cellist and an early member of the Electric Light Orchestra, is driving his van around the A381 in Halwell, Devon, United Kingdom. Some hundreds of meters up a hill, a 600 kg bale of hay falls from a tractor and starts rolling downhill. The timing between Mr. Edwards' van and the hay couldn't have been more perfect (or imperfect, depending on how you see it), and the massive bale crushes his van, killing him instantly. For Mike Edwards, there was no near miss. He "won" the anti-lottery. Had Mr. Edwards gone for another toast during breakfast or stopped to tie his shoelaces on his way out, he could have been telling this story over a pint in a pub.

Circa 2006, I am walking down Moreno Street in the Montserrat neighborhood, in downtown Buenos Aires. I am planning to have a sandwich for lunch, so I am heading to my favorite sandwich place at the intersection of Moreno and Lima streets, minding my own business, when I hear a loud sound right behind me. When I turn around, I see a small cloud of whitish dust. Once the dust settles, I see a big piece of concrete that has just fallen from above, only a few meters behind me. In fact, it is half of a balcony from three storeys up that has plummeted. Had I left my workplace 5 or 6 seconds earlier, that concrete piece would have landed on my skull. And I probably wouldn't be writing these lines right now.

Now, was that a near miss? In a way, it was. But, in another way, I had no control over it, just like Mike Edwards did not have any control over the rolling bale that ended his existence. How could I know the decay of the materials that made that

[3] It's on YouTube, go check it out, but please fight the algorithm and come back to me.

100-year-old balcony would give up at that exact time? I had no control over any variable in this situation. In fact, there might have been other near misses of such kind in my life that I am completely unaware of.

Near misses have a different taste when it's a combination of deterministic and stochastic processes. The Czech guy watching the building being demolished could have stayed home watching a movie, but he decided to go to a place where explosive charges would be used to bring a massive building down. Still on the very deterministic side of things, he and others purposely decided to stay at an insufficient distance from the events. He consciously put himself in the hands of probabilities. The fact that his head is still in one piece is the stochastic—read, lucky—combination of his position and the dynamics of the collapsing building, which created the piece of debris that could have decapitated him but didn't.

Near misses are puzzling because life appears to continue just like nothing has happened. From a certain perspective, nothing has in fact happened: there are no injuries, no blood, no hospitals or funeral houses involved. Still, all the mechanics of an accident were at play. Risks materialized to the fullest, and it was just a matter of chance at the end of the chain that defined whether we are talking about people getting hurt or not. Near misses are real-life, full-scale rehearsals of accidents.

Near misses are—or at least should be—transformative if you happen to suffer one. They broadcast a sense of fragility and act as a reminder of how the line between normal life and disaster is thin and gets redrawn all the time. Near misses force us to reflect on how lucky we were this time, and how we must be more attentive in the future on the things we *can* control if we want to prevent them from happening again.

Near misses happen in the engineering domain as well. Lots of them. Things that worked out barely and due to a combination of lucky events; things that could have ended in calamity should the planets have aligned in a slightly different way. In any case, it is always puzzling to see engineers adopting the mystic interpretation of a near miss, thinking that "what doesn't kill you makes you stronger", and believing an "angel guardian" or "good karma" protected them. Those who believe that are protected by holy spirits tell the stories of near misses in an epic way, completely missing the opportunity to learn and realize how close they were from disaster.

Chapter 31
The Office Conga Line

Abstract There are as many types of social awkwardness as there are engineers. Are we the weird ones, or are certain things in the workplace that weird us out? The COVID-19 pandemic might have slightly changed the way we interact with our colleagues, but things aren't too different; we still need to see the CFO doing the Macarena from time to time. And yes, we are fun at parties.

Keywords Team dynamics · Psychological safety · Workplace drama · Management

The stereotype says that engineers are socially awkward. Reality is that we are a bit of a weird bunch. We have a language and thought processes that can baffle non-engineering people. For the sake of avoiding falling into over-generalizations, I will today speak for myself and how such "social awkwardness" might have formed, shaped by weird situations around the workplace that molded the way I interact with colleagues.

I haven't been socially awkward all my life. I mean, without having ever been the life and soul of the party, I could say my childhood was adequate, socially speaking. Friends, football, music. The usual. By the time I started working—at 21—and while studying Electrical Engineering, I felt I was equipped with the basic tools to socially engage at the workplace. Or so I thought.

I have to admit, more than 20 years through many different professional environments took some toll. Small companies, big companies, more successful, less successful, more international, less international. Family businesses, multinational corporations, and startups. In my home country or abroad. I have seen a lot, and that lot has moulded me.

When I started, I was really into it. You know, getting to know people, sharing jokes, how was your weekend, how were the holidays, this and that. Hanging out with the office folks after work hours. Lots of small talk but also a genuine curiosity about how others were doing, what they liked, etc.

My first social questionings started with company parties. Initially, I thought ok, why not? Celebrate milestones, or the end of the year, whatever. Bosses will be

I. Chechile, *Engineering is Broken*, https://doi.org/10.1007/978-3-032-18169-5_31

there, kind of yikes, but who cares, we eat something, we have some fun, then I go home. That was the expectation at least. In reality, I am standing there, drink in hand, seeing Pedro from Accounting doing the limbo dance at the sound of Venus from Bananarama, tie in the head, Rambo-style, all sweaty, screaming unintelligible things in his native language.

What I started to see was that, for some of my workmates, the line between being a colleague to discuss requirements or a partner for a lambada train largely depended on the amount of margaritas consumed. And it's not that I didn't drink at all, and I only pointed a finger at sinful party animals who were simply having fun as if I were some sort of moral police. I can have a drink or two, but no amount of tequila can change the fact that Pedro the Accountant will always be Pedro the Accountant to me. The most unsettling part? The Mondays after these parties were quite odd for me, although everyone was acting like nothing had happened, except for Pedro, who, of course, would call in sick.

Then, as it happens to many engineers around the world, my career took me into managerial waters, and managerial waters are very socially demanding waters. As a manager, you find yourself leading people, so you are supposed to, well, at least get along, motivate, and guide them. You have to communicate a lot with them, sound reassuring, inspirational, and confident. There was a need for me to grow a mature, social, Zen-like emotional intelligence out of nowhere.

Unlucky as I tend to be, I did this transition at the worst context possible: an early-stage company where the HR department was still nonexistent, so I was basically left to my own devices when it came to dealing with people's issues. And believe me, people do have a lot of issues. Insecurities, unmet career expectations, egos, jealousies, mistrust, and personal tragedies.

How do you grow a Zen attitude in such a short time? You don't. You just survive as you try to understand what the hell is going on. And this is something leadership courses can't teach. I mean, some can help to certain extent, but you can't rely on a bunch of PowerPoint slides or what some random leadership coach told you when someone is opening up to you because they feel miserable. You can't just throw the platitudes those coaches spout, hand in chin, like: "Let's think outside of the box!". Leadership coaches do not really know what managing really means in a typical day, or they have forgotten after years of not doing it firsthand and spouting business babble for a living. There's a non-glamorous side of managing that requires industrial amounts of listening to people's fears and worries, and then—perhaps most importantly—doing something about it.

People's fears and worries vary a lot. Some put a lot of gravitas into things someone else would consider trivial. Once, someone came to my desk with a worried face and said, "I need to talk to you, it's urgent". I, a forming manager trying to learn to gauge body language after watching a few episodes of *Lie to Me*, saw a genuine expression of concern in this person's face, so I thought: Heck, this guy is quitting. So, he booked a meeting, and with the same funeral face, he said: "My laptop is too heavy, my back hurts, I need a lighter laptop". For me, it was a trivial problem he could've brought directly to IT. For him, it was something important that was clearly bothering him. And that gap between what a manager thinks it's important versus what their subordinate thinks is important can really define their relationship. From a manager's perspective, it feels as if your typical day is about being continuously interrupted for trivialities. For the ones interrupting you, their matters are far from trivial, and they feel they need immediate attention.

Infatuations and romantic shenanigans in the office have always been particularly hard for me to grasp. I mean, sure, people fall in love with each other; it happens. But the number of professional adults confusing their workplaces with a high-school prom is quite shocking.

People love to complain about work-life balance, but many voluntarily choose to make work their main social environment by making their lives so centered on the workplace that everything revolves around it. This is something frequently seen in very international companies like startups, where the combination of youth and being an expat away from home turns the companies into emotional shelters where people not only work but also find social containment. There's nothing inherently wrong with this, as long as the drama is kept below Dawson's Creek levels.

Harassment keeps happening, and nothing indicates it will go away. See this: Years ago, a manager of a division (so, my boss's boss) organized a surprise party for a secretary he was kind of obsessed with. He literally moved hundreds of people to the company's restaurant to lurk behind a wall to surprise this person. I could see the cringe in my poor boss's eyes as he was clearly forced to come and ask us to leave whatever we were doing and join the crowd. I managed to stay in my office in silent protest, thinking: man, this can't be happening, we have so much work to do, and we are throwing surprise parties? Yes, as you can see, I am fun at parties. Years later, I coincidentally shared office space with the person this guy was obsessed with, who commented on how much that party ruined their existence because everybody started to think they had special privileges.

So yes, emotions are always present in the office space, and still, I find engineers who think they will be able to strictly focus on technical matters all their lives: oh, sweet summer children.

There was a time when two subordinates had a nasty fight. There was a nascent HR department in place, so I went to them for some guidance. Here's the thing about HR departments in general: they have this tendency of making it look like they really, really care about the situation you're bringing to them, although you can see their eyes are actually dead and they are basically trying to get you out of their office as fast as they can. So, HR suggested a psychologist to decompress. I said, alright, why not? A psychologist was summoned sometime after. He looked more like a backpacker just arrived from a trip to Norway than a psychologist, but well, looks can mislead, and I don't know what I was expecting, honestly. Turns out, the belligerents refused to talk to him—it was not mandatory, I learned at that moment—so I ended up having to entertain the psychologist for an eternity as he was trying to analyze *me* instead.

It was not a total waste of time after all. He was a psychoanalyst, so thanks to him I got to know that I am in love with my mother, and that my family is full of assholes.[1]

There have also been unethical, borderline illegal situations I have been through. For example, ages ago, this former boss of mine asked me to help him install a keylogger into a (female) employee's computer he suspected was sharing "trade secrets" with a competitor. This was only his contrived plot to disguise that he wanted to spy on her, as he was inappropriately interested in this person. Of course, I refused. Shortly after, this female employee came to the IT guy—who was coincidentally sitting next to me—complaining that her computer was "acting weird". Without fully opening up to the IT dude (because I had no strong evidence that a keylogger had actually been installed in the end), I convinced the IT dude to change the poor woman's laptop for good. Why on earth was I dragged into all that noise? The more I saw those kinds of things, the more I wanted to hug my oscilloscope.

There are as many types of social awkwardness or anxiety as there are engineers. The COVID-19 pandemic might have slightly changed the way we interact with our colleagues, but things aren't too different. We still need to see things like the CEO doing the Macarena from time to time. Sure, not all engineers are social introverts; many will choose to join the lambada train and have a great time. Good for them. I like to remember that the workplace is not a high-school prom, and that we get paid to get shit done. And yes, I am a social butterfly.

[1] Disclaimer: I, by no means, am trying to turn the powerful Global Union of Psychoanalysts against me. I respect your work. Keep on reaching for that rainbow.

Chapter 32
Projects, Damn Lies, and Statistics

Abstract In engineering projects, we dance our way around uncomfortable truths rather than learning ways of confronting them like adults. With estimates that everyone knows are too optimistic, risk assessments that seldom include the real risks, and progress reports painted in yellow and green, project managers present schedules they don't believe in to bosses who know those plans are full of it, and everyone nods and moves forward with the farce as long as the customer is okay with it. In any other context, promising what you can't deliver has its own term: fraud. However, this is a term that does not appear in every engineering project's dictionary.

Keywords Project management · Statistics · Fat-tailed distributions · Probability distributions · Normal distribution · Gaussian statistics · Project planning

Way before mathematics was invented by humans, many processes were still normally distributed, even though we didn't have a name for it. The phenomenon existed independently of our ability to describe it.

The "bell" shape of the normal—also called Gaussian—distribution can be understood through the concept of aggregate effects of numerous small, independent factors. When we measure something, countless tiny, independent factors influence the result. These factors can include slight variations in the measurement tool, environmental conditions, or inherent variability in the thing being measured. Most of these small factors don't push the measurement extremely far from the average; they cause only small deviations. This is why most measurements cluster around a central value, called the mean. There's usually no inherent bias toward overestimation or underestimation in these. This symmetry in variation leads to a symmetrical distribution of measurements around the mean. As we move further from this central value, the probability of a measurement occurring there decreases. It's less likely that many factors will all align to produce a measurement much larger or much smaller than the average.

A great tool that helps visualize why this bell shape arises is the Galton Board. In this device, tiny balls drop through a series of pins, and balls fall into many parallel lanes. Each pin the ball hits knocks it slightly left or right. The numerous, small,

I. Chechile, *Engineering is Broken*, https://doi.org/10.1007/978-3-032-18169-5_32

independent deflections result in most balls ending up near the center lane, with fewer and fewer reaching the lanes on the extreme sides. This process creates a bell-shaped distribution at the bottom. As you turn the board around time and again, the overall distribution of the balls will be highly predictable every time, although where each and individual ball will end up will not.

The explanation of why we can predict how the tiny balls will distribute in such a confident manner is quite intuitive. When a ball hits a peg, there's a 50 %chance for it to go left or right. For it to fall in the leftmost slot, it would have to go left every time. For it to fall in the middle, it has to go left and right the same number of times. The most likely scenario is that a ball will hit some pegs going left and some going right without particular preference, so more balls end up in the center than on the edges. This creates a predictable distribution pattern. The bell shape is therefore a natural outcome of the aggregation of many small, independent influences, each causing slight deviations from an average, with no bias toward one direction or another. This aggregation naturally leads to more occurrences near the mean and fewer as we move away from it, creating the characteristic curve. In some fancier Galton boards, they add a few balls painted in a different color. If you visually choose one of those colored balls after they happened to land in one extreme, either left or right, you will see how in the next round that ball will surely rush back to join the average lot (the center lanes), which is nothing else than the known effect of regression to the mean.

Engineers, despite their supposed quantitative rigor and love of data, often misunderstand statistics when it comes to project-based engineering. We take the normal distribution— which works wonders for measuring tolerances of screws or signal noise—and blindly apply it to things like task estimation and project timelines where it has no business there. Then we act shocked when our carefully calculated estimates and their margins turn out to be wrong, when a six-month project takes two years, and we can't explain why. The reason is that we did the math correctly for the wrong underlying distribution, which is worse than not doing any math at all because it gives us false confidence in predictions that are fundamentally useless; the worst kind of wrong is being confidently wrong.

The problem begins with task estimation, which most project people treat as if it were a Mediocristan problem when it's actually an Extremistan one. Nicholas Taleb, in "The Black Swan" and "Fooled by Randomness", distinguishes between these two domains of randomness. Mediocristan is the province of the normal distribution, where outcomes are constrained, where no single observation can meaningfully change the average, and where extremes are rare. Human height lives in Mediocristan. You will never meet someone who is ten times taller than average and measuring a thousand people gives you a reliable estimate of the average height. Extremistan is the place of power laws and fat-tailed distributions, where single observations can dominate all others, where the average is meaningless, and where extremes are not rare outliers. Wealth distribution lives in Extremistan. One person can have more money than the bottom fifty percent combined, and measuring a thousand people tells you almost nothing about the actual distribution because the next person you measure might have more wealth than all previous measurements combined.

Engineering projects live in Extremistan, but we treat them with Mediocristan tools and then wonder why our forecasts don't work.

When we estimate a technical task, we think we're doing something like measuring heights. We look at historical data, we calculate an average, and we add some margin to be safe. In fact, by doing that, we equate implementing the task with dropping a ball through a Galton Board, where the outcome is determined by many independent factors, each pushing the duration slightly up or down in a symmetrical, bounded way. The task duration is not determined by many small independent factors. It's determined by a myriad of interdependent factors and ambiguities that can show up in the development process. Any one of these factors can dominate the entire task duration. These aren't small, independent, symmetrical variations around a mean. When a task takes ten times longer than estimated, we tend to think of it as an "outlier" and flag it as an anomaly, a rare event that we couldn't have predicted, some bad luck that fell outside our confidence intervals. This is the typical language of Mediocristan, where outliers are infrequent. In Extremistan, extreme events occur with sufficient probability that you should expect to encounter them regularly, not as shocking anomalies but as consequences of operating in a fat-tailed system.

The problem only gets worse when we combine task estimates into subsystem estimates and subsystem estimates into system estimates, adding up durations and margins as if we were adding independent normally distributed random variables. This would work if tasks were independent and actually normally distributed. But tasks have dependencies. A problem in one task spills into others. The testing task can't start until the implementation task finishes, and if the implementation shows that the requirements were wrong, you're screwed. But wait, because our stupidity sometimes meets no boundaries. We often treat schedule risk as if it were symmetrical. We calculate that a project will take six months "plus or minus" something, as if there's an equal probability of finishing early or late. The distribution is not symmetrical in the slightest, which means it's overwhelmingly more probable to be late than early.

When we create work breakdown structures, we decompose systems into subsystems and subsystems into components and components into tasks, and then we estimate each task, sum everything up, apply some margin derived from historical variance, and present a project schedule with fanfare. This entire structure rests on the assumption that aggregation will smooth out the variance, which is not true. The work breakdown structure creates an illusion of independence where there's none. The tasks are coupled through shared resources, technical dependencies, and information flow. A problem anywhere can propagate everywhere. The margin, calculated from historical knowledge, may capture the variations seen before, but will miss the fat-tailed nature of the distribution where the worst case is orders of magnitude worse than anything in our historical data.

Why do you think integration and verification phases consistently consume far more time than estimated, despite being rather understood activities done time and again on every project? Because integration is where all the optimistic assumptions embedded in the component-level estimates hit reality. Integration is where you discover that optimism and stupidity are practically synonyms in engineering. That there are mismatches and missing functionality and incompatible assumptions in the components that all tested fine in isolation. You discover that independent effects were only an illusion while the stuff was resting on the table alone but ceases to be as long as you put it next to the other stuff. In complex systems engineering, there aren't practically any independent effects, but correlated features and failures where one interface mismatch reveals a fundamental misunderstanding that affects multiple subsystems, where one missing requirement triggers a cascade of changes across components, and where one assumption being wrong means rearchitecting significant portions of the system.

The verification phase suffers similarly because we estimate it as if finding and fixing defects is a small, independent issue that takes some average time to resolve with some bounded variance. Some bugs are like that, but most aren't. Some bugs, especially in complex systems, may be almost impossible to reproduce or flag deeper problems and fixing them may take weeks or months of rework. You can't estimate this with a normal distribution mindset because you don't know which bugs will be simple and which will be heisenbugs[1] until you find them, and the tricky ones dominate the schedule far more than the simple ones affect it.

As "outliers" keep happening, common sense would indicate we should start noticing that our statistical thinking is wrong, but what do we usually do? We try to explain away specific cases rather than updating our thinking at large. Taleb calls this the "narrative fallacy", which is our tendency to construct stories that explain past events in ways that make them seem predictable in retrospect while doing nothing to improve our actual predictive ability going forward.

Engineering projects live in Extremistan, so we must abandon normal distribution thinking entirely. Leave it for measuring screws and washers. We need to estimate differently. Instead of point estimates with margins, we must get more used to using ranges that actually reflect the fat-tailed nature of the distribution. A task can't be forty hours "plus or minus ten hours". It's forty hours if everything goes way too well, two hundred hours if we hit complications, and potentially thousands of hours if we hit an architectural problem.

We need to plan differently. We must stop hoping for the best and have backup approaches for critical dependencies. We need to keep architectural decisions reversible as long as possible. We must factor in rework as default, not as contingency.

We need to interpret data differently, and stop treating outliers as anomalies to be explained away and start treating them as the most important data points one can have, the ones that reveal the distribution we're dealing with. When a task takes ten times longer than expected, we must stop blaming the engineers' competence or whatever narrative makes that occurrence feel more predictable or make us feel less accountable as managers. We must accept that our estimation approach is not accounting for the possibility of shitshow-level outcomes, which occur with surprising regularity. Almost every project I've been part of has had a shitshow moment. I bet the same applies in your experience. That doesn't sound like a "rarity", does it? Shitshows are not uncommon because our realities move in fat-failed probabilities.

[1] In computer programming jargon, a *heisenbug* is a software bug that seems to disappear or alter its behavior when one attempts to study it.

All this requires communicating differently with stakeholders, and trashing schedules with false precision and estimates that imply certainty about outcomes in fundamentally uncertain domains. Being honest about risk and uncertainty. We have been educated to see certainty as good and uncertainty as bad. It's actually the opposite. False certainty derived from poor statistics can be incredibly damaging. Honest uncertainty that reflects the nature of the problem at hand shows knowledge about what you're dealing with.

And the interesting part is that, when our flawed models inevitably go off, we don't take responsibility or revise our statistical assumptions. Instead, we do something far more interesting: we create new, fake schedules designed to tell stakeholders what they want to hear rather than what's actually happening or going to happen. We swap one piece of fiction for another, slightly more palatable piece of fiction.

32.1 Double Schedule, Double Trouble

In too many engineering organizations, two timelines tend to run in parallel. One is the customer-facing schedule: doctored, polished, happy, impossibly optimistic. The other is the internal schedule: ugly, real, delayed, not fun. This weird practice emerges as a very stupid organizational reflex to reconcile poor planning and optimism with reality, very often with bizarre consequences.

It starts somewhat innocently. Early in a program, enthusiasm is high, and every milestone looks achievable. The external schedule is made ambitious; not necessarily dishonest, just "stretching" a little to win the bid, to reassure some boss, or to show muscle to the customer. Internally, it looks a bit differently, and engineers start covering themselves and commenting on the things that could go wrong: supply chain delays, integration surprises. Over time, these two versions drift apart.

Projects do what projects do: they slip to the right. A critical component arrives late. A subsystem fails an acceptance test. A software bug plays hard to catch. A regulatory approval takes weeks longer than expected. Internal teams adjust their real schedule as they go, trying to maintain a fragile connection with the external, which still feels recoverable. Everyone wishes they can catch up or work overtime a little bit if needed. This, of course, never happens. And thus, all those pseudo-harmless things are not added to the external schedule, and it all feels like a small white lie. Who cares?

And so, it begins.

Many organizations operate in a way that bad news is punished, where admitting that a milestone will not be met and will therefore move is seen as a weakness. Teams fear the loss of confidence from customers, paving the way for dreaded things like the opening of contract renegotiations, penalties, or internal blame games. So, they hold the façade for as long as they can, hoping for a miracle. It's quite amazing how many engineering organizations' main strategy is based on miracles.

The double schedule is a strategy that sometimes buys a few months, but it almost always ends in an abrupt, painful, and at times comical disclosure.

The paradox is that technical customers, especially experienced ones, usually know something is wrong before they're told. They read the subtle and not-so-subtle signs: meeting cancellations or serial rescheduling, vague updates, or the sudden use of terms such as "re-scoping" or "redefinition". Insisting on keeping those fake schedules going damages trust faster than any schedule slip could. By the time the truth surfaces, trust has already been destroyed, and every further statement is treated with suspicion.

This dual-schedule behavior forces internal teams to work in a state of pretense, where they must appear to be on track in meetings and interactions with stakeholders while fighting with reality as soon as they go back to their desks. This creates a weird atmosphere, where everyone knows the truth, it floats around at all times, but no one dares say it. Every now and then, some innocent, pure soul might inadvertently "spill the beans", only to disappear forever from the face of the earth because of their inconvenient honesty.

Honest schedules don't make problems and delays disappear. However, when customers and stakeholders are treated as partners rather than adversaries, everyone can plan contingencies like grown-ups, adjust budgets, and align dependencies. Transparency builds trust, a currency far more valuable than a few months of artificial optimism. Running two schedules is a short-term protection that leads to long-term damage, and it has a practically negative cost-to-benefit ratio.

It's understood, it's usually the bosses' decision to pull off the double-schedule charade, and there's little you can do from the trench as a foot soldier. So, in case a boss is reading these lines while polishing their nth fake schedule for a review meeting with the customer: drop it, truth will out anyway. Come clean.

Chapter 33
Stop Making Idiots Rich

Abstract Engineers carry most of the risks that keep companies alive while absorbing the physical, mental and personal costs of designing and building things that work. However, the spotlight and the rewards tend to flow elsewhere, keeping the engineers fighting for a decent salary and better work conditions. This short chapter dives into the twisted dynamics between those who create real economic value and those who enjoy it, and how pride, discipline, and a sense of craft keep engineers returning to their desks every day to do their work despite everything. Far from a call to rebellion, this chapter intends to act as a friendly reminder: choose carefully who you want to make rich.

Keywords Value · Entrepreneurship · Risk

Engineers stay late debugging at 2 AM or giving support to the ops team during a release. They spend weekends optimizing a data processing function that shaves precious milliseconds off response times. They worsen their scoliosis every day a little bit further, hunched over mechanical keyboards in old, creaky chairs.

And still, the financial benefits flow in the direction of those who were in the room when the cake was cut, and the story usually goes that they deserve it because they had "the vision", when "the vision" was probably some childish diagrams they drew on a whiteboard and some slides with awful graphic design they crafted to convince some clueless people with money that those boxes and arrows represented something big and urgent. They are praised for their risk-taking, which in fact consists of fully borrowed risk: other people's resources, time, mental health, and reputations.

The engineer, meanwhile, takes real risks: physical risks (back issues, sight issues, just to name a few), the risk of their marriage ending because they neglect spouses or kids for a product launch, late meetings or last-minute work trips, the risk of not being next to their close ones because they're working abroad, or the risk of burning out so badly that they can't remember why they loved designing and building things in the first place. However, these risks don't come with the same rewards. They come with salaries that are always running behind something (inflation, industry's average, etc.), stock options that vest over many years and might end up leaving them

I. Chechile, *Engineering is Broken*, https://doi.org/10.1007/978-3-032-18169-5_33

empty-handed due to legal black magic (liquidation preferences, stock classes), and taxes.

However, the engineer builds the thing that sells and creates revenue. The engineer fixes the problem that saves a million dollars and creates the future products that will keep the company competitive. All while they read in press releases and social media posts that founders are the "visionaries" and "disruptors". In the meantime, another email lands from a higher up asking about engineering hours estimation, reminding them that they're simply a resource. Like oil or lumber.

Engineers know they're being underpaid relative to the value they create. The founder knows it too. The venture capitalists know. Everybody knows.

And yet, the engineer returns to their desk every morning, opens their shitty software tools, and begins work on the things that will make someone else rich. They'll do it brilliantly because they are people who take pride in their work. They'll stay late to get it right. But somewhere, in a nicer office with a door that closes and chairs that don't squeak, someone who can't code and struggles with basic grammar is sending an email about "synergy", "traction", and "strategy", slobbering an exit built on the backs of people who understand how to build things that work.

The system is somewhat perverse, but it is what it is. This is not a call for a revolt, nor some armchair socialist manifesto asking for the means of production to be in the hands of the proletariat. No. Do your work and do it right. Be a good employee. There's nothing especially wrong with contributing to creating shareholder value. But remember that you have the power to make people rich, so make sure you grant that wish to someone who deserves it, someone who knows what they're talking about, or at least cares to learn, and more fundamentally that cares at least a bit about you. Choose carefully; don't make stupid people rich.

Chapter 34
What the Future Holds: Engineering, AI, and the Machine That Designs the Machine

Abstract The prospect of an artificial intelligence entity capable of designing and constructing complex physical systems has captured attention lately, often framed as a scenario when human engineering becomes obsolete in the hands of these futuristic constructs. For the hype masters, it is easy to imagine digital superintelligence conceiving a machine, generating its schematics, orchestrating its manufacturing, and assembling the final product in a seamless, automated chain. However, this vision assumes that engineering is a tidy, fully digital discipline without gaps across domains. What would an algorithm need to replace us?

Keywords Artificial intelligence · AGI · Digital economy · Digital engineering · Automation · Software development

As they say, making predictions is difficult, especially about the future. This is a chapter that felt risky to write. Because I have no crystal ball, things may go down in a way that what I write below might age poorly. I'll take the risk.

Throughout this book, we explored the many inefficiencies that surround our craft. Being a fatally human activity, it shouldn't come as a surprise that we humans are an essential ingredient of all these pitfalls in the practice when we let our insecurities, fears, and emotional traits shape the way we engineer.

The title of the book claims that engineering is broken. If that's the thesis, then how do we fix it? The Vasa sank in 1628 due to factors that are still very present in everyday life in the engineering workplace, so it seems the problems have been there for quite a long time. If nothing is done, the issues will remain for centuries to come, long after I and the ones reading these lines will be pushing up the daisies.

Is this time different, then? Do we fix it with AI? Could an algorithm keep our future figurative Vasas afloat?

AI (note I will use the umbrella term as everybody does, although AI means nothing by itself, and it's a collection of many subdomains with specific applications) has been making undeniable progress in the last few years, and it seems to be accelerating in its capabilities. We are still in the phase when we're understanding

I. Chechile, *Engineering is Broken*, https://doi.org/10.1007/978-3-032-18169-5_34

how to use it in our profession, and the risk here is that we will use it to also automate what is wrong with what we do. You can write terrible requirements with AI, because requirement quality depends on capturing what your customer wants, and if the customer is a human, I struggle to see how an AI will understand a human better than another human. It feels as if the Vasa could have still sunk with advanced AI because the problem was the King, his royal stubbornness, and the fact that no one dared to say no. Could an AI say no to the kings of the future and curb their random caprices?

The prospect of an artificial intelligence entity capable of designing and constructing complex physical systems is often framed as a scenario in which human engineering becomes obsolete in the hands of these futuristic things. For the hype masters, it is really tempting to imagine such a digital intelligence conceiving a machine, generating its schematics, orchestrating its fabrication, and assembling the final product in a seamless, fully automated chain. But this vision assumes that engineering is a tidy, truly digitalized discipline without gaps across domains, when, in reality, engineering is a messy, disjointed ecosystem, deeply entangled with bad tools, brittle supply chains, bottlenecks, and logistical dependencies that no cognitive superintelligence will have it easy handling.

A sufficiently advanced AI could, in principle, derive the physics of a device, optimize its structure, and produce the most elegant and efficient design ever conceived. However, the moment it tried to move from theory to implementation, it would collide with a world fundamentally made for humans and by humans. Today's design environment is not a computational substrate, but a convoluted landscape of proprietary tools that crash at random times, suppliers' websites, repositories, broken APIs, and toolchains shaped either by market forces or by a random dude in Nebraska.[1] Nothing about this environment resembles the ascetic terrain that artificial general intelligence scenarios rely on.

Even if this hypothetical super-capable AI managed to navigate the disjointed nature of the design process, for it to cross the gap from the digital to the physical, it would then confront the badly behaved arena of manufacturing. Industrial systems are built around human operators, human-readable processes, and human organizational rhythms. Supply chains run on elusive pricing, non-disclosure agreements, pointless secrecy, purchase orders, inventory systems, and customs paperwork. Assembly lines respond to tolerances, certification processes that move at snail speed, and machine interfaces designed for featherless bipeds. The manufacturing world is unmistakably analog, full of quirks, bottlenecks, and non-linear dependencies.

This is where the fantasy of a machine seamlessly designing and building machines starts losing some of its screws. Digital superintelligence alone cannot produce a circuit board if the fabrication house will reject the stack-up at least twice. It cannot accelerate the curing time of adhesives and coatings, or the shipping time of freight crossing ports operated by grumpy customs agents. It would require the whole shipping industry to go robotic, which might be the direction it's slowly moving to,

---

[1] Obligatory XKCD reference: "Dependency", comic nr 2347.

but still far, far from having the level of automation necessary to be handled by some algorithm through API calls.

All in all, the idea of an artificial intelligence effortlessly producing complex machinery reflects an assumption that everything that matters lives inside the digital realm. However, the true difficulty of engineering is in everything outside of it. The physics, the CAD, the simulation, the optimization, all of these are the "easy" parts. The hard parts are the interface delays, the crosstalk in the boards, or customs holding a shipment for fifteen days because a form was filled incorrectly, and the heisenbug that hides behind three obscure registers in a 12,000-page technical manual or an errata published inconspicuously at the bottom corner of a website.

An AGI could master the theory of every engineering domain, yet still be defeated by the practicalities that glue them together. The world is not a continuous computational space. It is full of gaps where the biggest pains lurk. And as long as the physical, logistical, and organizational machinery of engineering remains built for and around humans, any intelligence that hopes to shape the physical world will be forced to navigate the same constraints we do.

This is not an argument that AI will have little impact; on the contrary, it will transform design and creative workflows deeply, and it's already doing it. Software development will be changed forever; in fact, it might be seeing its last days as we know it, where writing code by hand will be more associated with craftmanship[2] than anything else.[3] Also manufacturing. But this is good for engineering. We need that. We need to stop relying in tribal knowledge, meetings, incredibly manual tasks, and graybeards who skimp on their knowledge and experience for their own job security. We need a less analog engineering. We need machines to take over, because corporate organizations are borderline unfeasible as they are, and perhaps this is the technology we have been needing to make them better. However, the belief that artificial intelligence alone can erase the brittle human scaffolding of engineering is a mistake born of wishful thinking from those hyping from mediocre articles and white papers in business quarterlies who have never built anything themselves. The biggest challenge will be to encode the physical world into computation. And until the entire industrial ecosystem becomes as machine-readable as the algorithms that aspire to control it, even the most advanced AI will remain trapped in the same dependencies, delays, and obscurities that shape engineering today.

---

[2] Among developers, one can find two kinds: those who love the code and those who love the product. For the code purists, AI is a curse because it skips the craft and the elegance. For the "product builder", AI is godsent because it represents a faster way to ship, at the risk of shipping garbage, of course.

[3] In all fairness, before AI, we were shamelessly copying code from everywhere, like Stack Overflow and other places, and no one batted an eye. Whoever says they were writing TCP clients or servers entirely by hand in C or Python is blatantly lying.

34.1 A Robot Stole My Job, and I Couldn't Be Happier

Ok, now I got a bit triggered, so this section goes a little bit beyond the engineering craft only, and it generalizes the points made before to the rest of our (still rather analog) lives. This is relevant because our professional practice is embedded in a broader context that is under transformation, and our activity is shaped by such contextual changes.

We've surely dragged our feet to the supermarket to mindlessly tour around the aisles to do our groceries, putting in the cart the usual stuff. Then, queueing in a long line while totally zoning out or browsing Reddit. And yes, home delivery of groceries is a thing; there are plenty of apps to order online and get the stuff delivered at your door. Is that what they mean by 'digitalization'? Are we fully digitalized, then? Hell nah. Digitalizing only the last stage—the customer-facing stage—of an otherwise incredibly analog, long chain is not what 'digitalizing' means. Adding nice-looking user-interfaces to the last bit does not replace the humongous choreography of human flesh involved in the process; it does not replace the person ringing your doorbell or leaving the stuff at your door, or replaces the person driving the van, or the person loading the van, or the people rushing around the warehouse taking things from shelves as per your order and putting them in a box. Nor does it replace the people who unloaded the trucks at the supermarket warehouse or those who populated the shelves, or the person driving the truck to the warehouse, or the people working at the factories that supplied the supermarket, where the problem replicates all over again. It is a largely manual, human-driven activity of moving stuff around, lifting things up and down, turning machines on and off, and a lot of people moving with their legs. The same goes for any other delivery service, such as your typical food delivery app, where sleek smartphone applications give customers the illusion of a more "digital" experience when ordering food where in reality there is a long list of low-wage workers in flesh and bones using their muscles to cook and ship your food, including the neglected, sweaty "partner"—you gotta love the cruelty of calling them that way—pedaling their life no matter the weather to get your food home before it gets cold, hoping for a good review.

Now think of the opposite; an hypothetical, truly digital scenario: you order your stuff with your phone, but instead of all the humans referred above, this time your order is truly processed by software which sweeps your list of desired products and tasks machines to bring your stuff from shelves which were populated by other machines as another software ensured beforehand that the shelves stuffed according historical trends in demand of the products, in a way the algorithm can place orders by commanding supplier-side software, which in turn tasks their factory accordingly, adjusting—and to some extent predicting—demand and also defining the routes and tasking of the delivery machines, optimizing for the shortest path and providing estimated time of arrival with good accuracy.

Now think about all the technology that must be in place for this full digital scenario to be possible: autonomous robots for loading/unloading lorries and organizing stuff in the warehouse while updating all related databases and systems.

And lots of software capable of processing time-dependent consumption trends and mining valuable data for tuning the ordering and manufacturing process.

Have you traveled abroad recently? That's some analog experience, isn't it? You need to drag your bags to the car, drive to the airport, manually deliver the bags, walk your soul through security, kill time in the duty-free shop, eyeball screens, freak out, run to the gate, and board the plane. Although it is hard to imagine a future of robots at home packing your underwear for you, it's still possible to think of a future where traveling involves autonomous vehicles picking you and your bags up from your door, connecting you to a bigger shuttle, ensuring your bag is routed to the right flight and destination without you having to roll it. Then, all you have to do is the physical security check. No screen-eyeballing (you would be informed of any change in real-time, or you could have the list of flights projected in smart glasses), no bag carrying, no more running, because you would have all the information to be on time. Then, at your destination, the counterpart of the digitalized transportation system gets you and your bag from the airport at the right time and hauls you where you wanted to be. Today, you even have to fight for electrical outlets to charge your phone at airports. Not a very "digital" experience, as much as the marketing departments of the air transportation industry want us to believe with their smooth-voiced ads.

From what I've stated above, it appears as if the "digital" future is about huge, pristine warehouses full of robots scrambling around, with no human presence at all. Or about airports without counters and without polite servers shouting 'next!'. It appears as if the digital economy means that all the people who is part of the analog choreography when we do the groceries would eventually lose their jobs. Is this true? It might be, and this, in fact, couldn't be better news, because it would be the chance for them to get better jobs.

As much as driving a truck is a respectable activity as any other, it does not cognitively exploit the best of what a human brain can do. A human sitting behind a wheel for hours and days with the sole intention of keeping the wheeled vehicle between two parallel lines and to avoid bumping into other vehicles to make it possible for the cargo to reach its destination sounds, if looked at from a reasonable distance, insulting for our evolution as a species. Same for using humans to unload a truck, or carry boxes around, or driving a forklift across a warehouse to move boxes. Or scan products through a laser beam for hours while listening to seniors competing on who's the sickest. Or checking a passport behind a counter. We are more sophisticated than that, and the fact that those jobs still exist is a reminder of the extremely analog society we still live in. Hopefully, robots and software will not destroy jobs irreversibly but incentivize the workforce to shift into more cognitively meaningful jobs because the 'digital' economy, as such, will need lots of thinking. Human thinking. Architects, designers, debuggers, configurators, maintainers, testers. Our human brains are capable of extremely complex, nuanced operations: from poetry, music, all the way to complex systems engineering. An unspoken side of inequality in the world is the cognitive difference between low-wage jobs and white-collar, better-paid jobs. The fact that today 'low wage' equals 'low cognitive intensity' should concern us more than it does.

If a serious digitalization of the economy were to ever get past the marketing buzz, all layers of the education system would need to adjust accordingly, tuning curricula for the skills and capabilities needed for an economy where humans driving a truck will not be needed anymore. Where humans stacking boxes in a warehouse will be remembered with humor, as knocker-uppers are remembered today. Instead of the sterile discussion of robots coming to steal our jobs, the discussion should be about how well prepared we are to speed up the process of robots taking our boring jobs, so we can go and do more meaningful stuff. It's an exciting time to be alive, and an exciting time to be an engineer.